Basal Reinforced Piled Embankments

The Design Guideline

Editors: Suzanne J.M. van Eekelen and Marijn H.A. Brugman

CRC Press is an imprint of the
Taylor & Francis Group, an **informa** business
A BALKEMA BOOK

Design Guideline Basal Reinforced Piled Embankments

Published by	CRC Press
E-mail	Pub.NL@taylorandfrancis.com
Editors	Suzanne J.M. van Eekelen
	Marijn H.A. Brugman
Book design by	Lydia Slappendel
Drawings made & provided by	Suzanne J.M. van Eekelen

www.crcpress.com – www.taylorandfrancis.com

ISBN: 9789053676240 (hbk)
ISBN: 9781032402451 (pbk)
ISBN: 9781315389806 (ebk)

CONTENTS

Preface

Publication 'CUR 226 Design Guideline for Basal Reinforced Piled Embankment Systems' ('CUR 226:2010') appeared in March 2010. More recent insight has made it necessary to amend 'CUR 226' on a number of points.

This design guideline ('CUR226:2016') has been completely revised and this document is a translation of the greater part of this revised text into English. The primary changes with respect to the 2010 version are:

- A new, improved design model has been incorporated for the design of the geosynthetic reinforcement: the Concentric Arches model (Chapter 4).
- For this new design model, a new set of factors has been determined: partial safety factors and a model factor (Chapters 2.6 and 2.7).
- The traffic loads have been adopted in accordance with the Eurocode (Chapter 2.3).
- Extensive calculation examples have been added for the design of the geosynthetic reinforcement (Chapter 5).
- The instructions for conducting the numerical calculations have been updated (Chapter 6).

The Dutch language version of this design guideline also contains extensive chapters about pile and pile cap design. These have been condensed for this English version because they are of less interest to the non-Dutch reader. Moreover, Chapters 7 to 9 have been condensed for this translation. With this revised version of the Basal Reinforced Piled Embankment Design Guideline, engineers have available a guideline that includes the state-of-the-art knowledge and experience with this type of embankment construction which has also been validated by means of practical experience and measurements.

The provisions in this design guideline are expressed in sentences in which the principal auxiliary verb is "should".

An excel file containing the basic equations of the Concentric Arches model is available for download at **www.piledembankments.com** or at **www.crcpress.com/9789053676240**.

We would appreciate your comments on, suggestions for or experience with this design guideline. Please share them with us at suzanne.vaneekelen@deltares.nl.

The editors

Summary

A basal reinforced piled embankment consists of a reinforced embankment on a pile foundation. The reinforcement consists of one or more horizontal layers of geosynthetic reinforcement installed at the base of the embankment.

A basal reinforced piled embankment can be used for the construction of a road or a railway when a traditional construction method would require too much construction time, affect vulnerable objects nearby or give too much residual settlement, making frequent maintenance necessary.

This publication is a guideline for the design of basal reinforced piled embankments. The guideline covers the following subjects: a survey of the requirements and the basic principles for the structure as a whole; some instructions for the pile foundation and the pile caps; design rules for the embankment with the basal geosynthetic reinforcement; extensive calculation examples; numerical modelling; construction details and management and maintenance of the piled embankment. The guideline includes many practical tips. The design guideline is based on state-of-the-art Dutch research, which was conducted in co-operation with many researchers from different countries.

In the preparation of this Basal Reinforced Piled Embankment Design Guideline, the composition of CUR Committee 1693 was as follows:

	Name	Organisation	Contribution
*	dr. Suzanne J.M. van Eekelen	Deltares	Chairperson and editor. Concentric Arches design model for the geosynthetic reinforcement, calculation examples.
*	Marijn H.A. Brugman	Arthe Civil & Structure	Editor.
	Joris van den Berg	Low & Bonar	
	Henkjan Beukema	Dutch Ministry of Public Works dept. for Road Construction	
	Diederick Bouwmeester	Ballast Nedam	
	Jeroen W. Dijkstra	Cofra	
*	Piet G. van Duijnen / Constant A.J.M. Brok	GeoTec Solutions / Huesker	Safety philosophy and model factor and partial factors.
	Dick W. Eerland	Eerland Bouwstoffen Management	
*	Jacques Geel	Heijmans Infra	Pile cap design and calculation examples (in the Dutch version of this guideline).
	Jos Jansen	Volker InfraDesign	
*	Martin de Kant	Royal HaskoningDHV	Fill properties.
	Gert Koldenhof	Citeko	
	Leo Kuljanski	Tensar International	
*	Maarten ter Linde	Strukton	
	Herman-Jaap Lodder	RPS Group Plc	Review of equations.
	Sander Nagtegaal	Voorbij Funderingstechniek	
*	Eelco Oskam	Movares	
	Tara C. van der Peet	Witteveen+Bos	Review of text.
*	Marco G.J.M. Peters	Sweco	Traffic load.
*	Maarten Profittlich	Fugro GeoServices	Calculation examples.
	Jeroen Ruiter	TenCate Geosynthetics	
*	Daan Vink / Bas Snijders	CRUX Engineering	Numerical, transition zones, stiff/non-stiff embankment.
*	Lars Vollmert	Naue / BBG	Adaptation κ model.
	Robbert Drieman / Fred Jonker	SBRCURnet	Project managers.

** Core group members*

For the purposes of the expert programme conducted for the guideline, financial and/or in kind contributions were received from:

» Arthe Civil & Structure
» Ballast Nedam
» BRBS Recycling (Sector organisation for sorting and dismantling)
» Citeko / NAUE / BBG
» CRUX Engineering
» Deltares
» Dutch Chapter of the IGS (NGO)
» Fugro GeoServices
» Huesker / GeoTec Solutions
» Low & Bonar
» Movares
» Rijkswaterstaat (Dutch Public Works) - Large Projects & Maintenance
» RPS Advies- en ingenieursbureau
» Stichting Fonds Collectieve Kennis - Civiele Techniek (FCK-CT) (Collective Knowledge Fund Foundation - Civil Engineering)
» Strukton
» Sweco
» TenCate
» Voorbij Funderingstechniek
» Deltares

SBRCURnet wishes to express its thanks to all the members of the Committee and their organisations, for the contributions made to this output from SBRCURnet Committee 1693, 'Revision of CUR 226 Design Guideline for Basal Reinforced Piled Embankments'.

1st version January 2010 in Dutch	CURNET management
2nd completely revised version 2016 in Dutch	SBRCURnet management
2nd completely revised version 2016 in English, first print	SBRCURnet management
2nd version, 2016, second print (2019)	CRC Press / Balkema

The English-language version of this design guideline has come into existence thanks to the financial and practical support of Deltares, TenCate, Naue and Arthe Civil & Structure.

Nomenclature

The most important parameters are:

A	kN/pile	Load part transferred directly to the pile ('arching A' in this guideline) expressed in kN/pile = kN/unit cell
A_L (A_{Lx}, A_{Ly})	m^2	GR area belonging to a GR strip in x or y direction respectively, assuming circular pile caps, see equations (4.25) and (4.26)
A_i	m^2	Area of influence of one pile unit, $A_i = s_x \cdot s_y$
A_g	m^2	Area of the geosynthetic reinforcement between the pile caps, $A_g = A_i - A_p$
A_p	m^2	Area of a pile cap
A_1, A_2, etc		Reduction factors according to EBGEO (2010) for the GR strength, see Chapter 2.10.1
$A\%$		See E below
B	kN/pile	Load part that passes through the geosynthetic reinforcement (GR) to the pile, expressed in kN/pile = kN/unit cell
b	m	Width of a square pile cap
b_{eq}	m	Equivalent width of a circular pile cap
C	kN/pile	Load part that is carried by the soft soil between the piles (this soft soil foundation is called 'subsoil' in this document) expressed in kN/pile = kN/unit cell
ctc		Centre-to-centre (spacing of piles)
d	m	Diameter of a circular pile (cap)
d		Design value (subscript)
d_{eq}	m	Equivalent diameter of a square pile cap
E or $A\%$		Pile efficacy or pile efficiency

$$E = A\% = 1 - \frac{B+C}{w_{tot}} = \frac{A}{w_{tot}} = \frac{A}{A+B+C}$$

E	kPa or MPa	Young's modulus
F	kN	Force
F_d		Factored design load
F_k		Unfactored characteristic load
f	H_z	Frequency
f or z_{max}	m	Maximum GR deflection
f_{m1}, f_{m2}, etc		Reduction factors according to BS8006 (2010) for the GR strength, see Chapter 2.10.1
GR		Geosynthetic Reinforcement
g	m	geosynthetic (subscript)
H	m	Embankment height, (between road surface and pile cap)
H_{eq}	m	Equivalent fill height for the determination of the normative traffic load, see Chapter 2.3.2
H_{g2D}	m	Height of the largest of the 2D arches in the new Concentric Arches model, see equations (4.19) H_{xg2D} refers to the height of a 2D arch that is oriented along the x-axis, H_{yg2D} refers to the height of a 2D arch that is oriented along the y-axis, as shown in Fig. 4.5
H_{g3D}	m	Height of the largest 3D hemisphere in the new Concentric Arches model, see equation (4.13) and Fig. 4.5
h^*	m	Fill height for which conditions apply for the fill strength (fill friction angle φ')
$J\ (J_x, J_y)$	kN/m	Tensile stiffness of the GR (GR stiffness)
K	kN/m³	Modified subgrade reaction value (see equations (4.29) and (4.30)). This value is used to account for 'all subsoil', not only the part below the GR strips between adjacent piles. K_x is for a GR strip oriented along the x-axis. K_y is for a GR strip oriented along the y-axis
K_a		Active principal stress ratio or active earth pressure coefficient $K_a = \frac{1-\sin\varphi}{1+\sin\varphi}$

Symbol	Unit	Description
$K_{p'}$, K_{crit}		Passive or critical earth pressure coefficient $K_p = K_{crit} = \tan^2\left(45^o + \frac{\varphi'}{2}\right) = \frac{1+\sin\varphi}{1-\sin\varphi}$
k		Characteristic value of a parameter (subscript)
k_s	kN/m^3	Subgrade reaction
L_w (L_{wx}, L_{wy})	m	The clear distance between adjacent pile caps ($L_{wx} = s_x - b_{eq}$ and $L_{wy} = s_y - b_{eq}$)
L_{x2D}	m	Part of the GR strip that is oriented along the x-axis and on which the 2D arches exert a force, see Fig. 5.4 and equation (4.18)
L_{y2D}	m	Part of the GR strip that is oriented along the y-axis and on which the 2D arches exert a force, see Fig. 5.4 and equation (4.18)
L_{3D}	m	Width of GR square on which the 3D hemispheres exert a load, see Fig. 5.3 and equation (4.14)
N	kN	Axial force
N		Number of heavy truck passages (> 100 kN) per year and per lane (NEN-EN 1991–2)
P_f		Risk of failure
P_{2D} (P_{x2D}, P_{y2D})	kPa/m^{Kp-1}	Calculation parameter in the Concentric Arches model, given by equation (4.21). P_{x2D} refers to a 2D arch that is oriented parallel to the x-axis, as indicated in Fig. 4.5. P_{y2D} refers to a 2D arch that is oriented parallel to the y-axis
P_{3D}	kPa/m^{2Kp-2}	Calculation parameter in the Concentric Arches model, given by equation (4.12)
p	kPa	Uniformly distributed surcharge on top of the fill (top load), spread to the GR level
p		Pile or pile cap (subscript)
$p_{permanent}$	kPa	Permanent surcharge on top of the fill, for example a load from an abutment
$p_{traffic}$	kPa	Design traffic load according to the Eurocode, as given in Table 2.2, Table 2.3, or Table A.1 t/m Table A.8. If the permanent load $p_{permanent} = 0$, we find that $p = p_{traffic}$

$p_{traffic;max}$	kPa	Maximum vertical pressure on the GR arising from the traffic load
PET		Polyester
PP		Polypropylene
PVA		Polyvinyl alcohol
Q_{2D}	kN/m^3	Calculation parameter in the Concentric Arches model given by equation (4.21)
Q_{3D}	kN/m^3	Calculation parameter in the Concentric Arches model given by equation (4.12)
q_{av}	kPa	Average load on GR strips, see equation (4.28)
R	m	Radius
RF_{creep}		Reduction factor for the loading time (creep) according to BS 8006 (2010)
s_d	m	The diagonal centre-to-centre distance between piles $s_d = \sqrt{s_x^2 + s_y^2}$
s_x, s_y	m	Centre-to-centre pile distance parallel to the *x*-axis or parallel to the *y*-axis
T_H	kN/m	Horizontal component of T_v
T_h	kN/m	GR tensile force due to the horizontal spreading force (lateral embankment thrust)
T_{max}	kN/m	Maximum value of T_v, at the edge of the pile cap
T_r	kN/m	GR tensile strength
T_s	kN/m	Total GR tensile force due to the vertical and the horizontal load
T_V	kN/m	Vertical component of T_v
T_v	kN/m	GR tensile force due to the vertical load
v		Vertical (subscript)
V	m^3 or m^3/m	Volume
X_d		Design soil parameter value
X_k		Characteristic soil parameter value

z	m	Depth in m
z_{max} or f	m	Maximum GR deflection
α	o	Angle
α	m^{-1}	Calculation parameter for calculation step 2, $\alpha^2 = K/T_H$
β	o	Angle
β		Reliability index
γ	kN/m^3	Unit weight
γ_f		Partial factor for load
γ_m		Partial factor for material behaviour
ε		GR strain
ε_{max}	kN/m	Maximum value of ε, at the edge of the pile cap
κ		Reduction factor to reduce arching due to cyclic loading according to Heitz (2006), see Fig. 4.6 and equation (4.10)
μ	-	Friction factor
ν		Poisson's ratio
$\sigma_{v;g}$	kPa	Average vertical pressure on the GR between the pile caps
$\sigma_{v;p}$	kPa	Average vertical pressure on the pile caps above the GR
$\sigma_{v;tot}$	kPa	Average vertical stress at the level of the bottom GR layer (neglecting arching)
φ'	o	Internal friction angle under effective stress conditions
φ_{cv}'	o	Critical state value of the internal friction angle under effective stress conditions
φ_p'	o	Peak value of the internal friction angle under effective stress conditions

1 Introduction

1.1 General

Worldwide, more and more basal reinforced piled embankments are being constructed for transport infrastructure. A basal reinforced piled embankment (Fig.1.1, Fig.1.2) consists of a reinforced embankment on a pile foundation. The reinforcement consists of one or more horizontal layers of geosynthetic reinforcement (GR) installed at the base of the embankment. In use, these structures exhibit little or no residual settlement.

The force transfer in the reinforced embankment is determined by arching. This is the phenomenon where loads are transferred preferentially to the stiffer elements in the ground, in this case the piles.

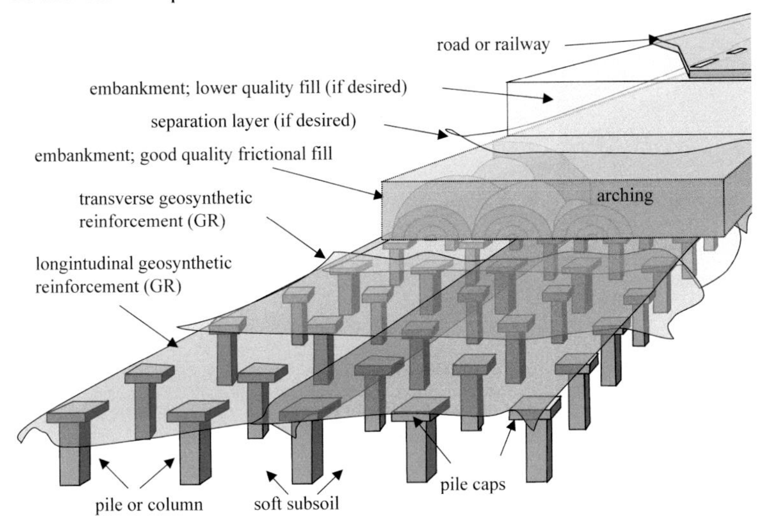

Fig.1.1 A basal reinforced piled embankment

The pile caps are preferably positioned with their tops above the groundwater level.

All possible pile systems may be used for piled embankments, providing that the difference in stiffness between the piles and the surrounding soil is sufficiently great; see Table 4.2, boundary condition 8. Important points in the structural design are the calculation of the piles' bearing capacity, for which the regulations in force for the design of piles are used, and the dimensioning calculation for the geosynthetic reinforcement itself.

Fig.1.2 Basal reinforced piled embankments under construction. *(a)* Krimpenerwaard N210 (Ballast Nedam, Huesker, Fugro, Movares), *(b)* A-15 MAVA project, source: Royal TenCate, contractor: A-Lanes *(c)* Piled embankment for an abutment necessary for the widening of the A2 near Beesd, the Netherlands (Voorbij Funderingstechniek, Heijmans, CRUX Engineering, Huesker and Deltares), *(d)* Houten railway (Movares, de Bataafse Alliantie, (ProRail, Mobilis, CFE en KWS Infra), Huesker, Voorbij Funderingstechniek, CRUX Engineering and Deltares), *(e)* Krimpenerwaard N210 (Ballast Nedam, Huesker, Fugro, Movares), *(f)* Hamburg (Naue).
Figure published before in van Eekelen (2015), [23].

The design process for a basal reinforced piled embankment proceeds as indicated in Table 1.1. The steps 1 to 3 may be seen as the preliminary design. In these steps, the pile arrangement including pile type and GR strength are determined. Step 4 may be seen as the final design, in which additional calculations are done to determine for example bending moments in the piles, with the help of numerical calculations.

The following are considered in this publication:

- requirements for the reinforced embankment;
- requirements for the piles and pile caps and recommendations for pile and pile cap design;
- design of the reinforced embankment, including calculation examples;
- evaluation of pile moments with numerical calculations (finite element method, FEM);
- transition zones;
- construction and maintenance of the piled embankment.

Table 1.1 Stages of piled embankment design.

Step	Designation	Parts	
1	Overall dimensions	a.	The geometry (thickness, width) of the embankment taking into account requirements from the surrounding area, freeboard and punching.
		b.	The fill material of the embankment is chosen; the various standards impose environmental and structural requirements on the fill.
2	Calculation of the bearing capacity of the **piles**	c.	Both geotechnical and structural; determination of the centre-to-centre spacing of the piles. See the considerations in Chapter 3. The piles are often the largest cost item of the structure. At the transition to a non-piled road section, the pile spacing is sometimes increased gradually and/or the pile toe depth is reduced.
3	Design of the reinforced **embankment**	d.	Design calculation for the geosynthetic reinforcement (GR). • Vertical load, based on the arching theory; this calculation is done using analytical formulae, see Chapter 4.3.2; • Horizontal load, due to vehicle braking, centrifugal forces; see Chapters 2.4 and 6.2.; • Horizontal load, due to lateral thrust in the slope; this calculation is done using analytical equations, see Chapter 4.3.3.
4	Checking the settlement and stability	e.	Expected settlement of the pile foundation, both geotechnical and structural.
		f.	Possible bending moment in the piles, using FEM calculations, see Chapter 6.
		g.	Total GR strain and in-service GR strain (due to traffic load and creep), or the expected settlement due to the GR strain.
		h.	Check of the overall stability.
		i.	Check of the subsoil support (if applicable). The settlement of the reinforced embankment and piles occurs fairly quickly after the application of load, depending on the rate at which the subsoil permits any deformation in the reinforcement. The overall settlement of a basal reinforced piled embankment is negligible, if it is designed and constructed correctly.

1.2 Eurocode

This 2016 publication update has been brought into conformity with the requirements of the European Eurocode. Table 1.2 gives an overview of the Eurocodes applied.

Table 1.2 Overview of Eurocodes*

Eurocode NEN-EN	Title
NEN-EN 1990+A1+A1/ C2:2011 [3]	National annex to NEN-EN 1990+A1+A1/C2: Eurocode: Basis of structural design
NEN-EN 1991-1-4+A1+C2:2011 [4]	National annex to NEN-EN 1991-1-4: Eurocode 1: Loads on structures
NEN-EN 1992-1-1+C2:2011/NB:2011 [5]	National annex to NEN-EN 1992-1-1+C2 Eurocode 2: Design of concrete structures - Part 1-1: General rules and rules for buildings
NEN 9997-1+C1:2012 [8]	Geotechnical design of structures - Part 1: General rules
NEN-EN-ISO 22477-1:2006 [9]	Geotechnical investigation and testing - Testing of geotechnical structures - Part 1: Pile load tests by static axial compression

** as valid and in force on date of this 2016 edition of this publication*

The European standards mainly concern the structural design of buildings. The requirements specified in them do not always apply to other civil engineering structures, such as embankments, bridges and viaducts.

2 Requirements and initial details of reinforced embankments

2.1 Introduction

The client's principal requirements for the construction of a piled embankment concern:

- deformations / differential deformations;
- external loads;
- reliability class;
- service life / need for maintenance;
- limitations in use and effect on immediate environment.

These requirements are covered in this chapter.

This chapter presents additional principles and details concerning:guideline.

- pile location tolerance;
- support of the geosynthetic reinforcement by the subsoil;
- surface water;
- applicability of this design.

Chapter 3 gives principles and details concerning piles and pile caps.

In a piled embankment, the vertical load is distributed over the piles, geosynthetic reinforcement and subsoil as follows (see Fig.2.1):

- the load part *A* ('arching *A*') that goes directly to the piles via the arching effect;
- the load part *B* that goes to the piles via the geosynthetic reinforcement;
- the load part *C* that is carried by the subsoil between the piles.

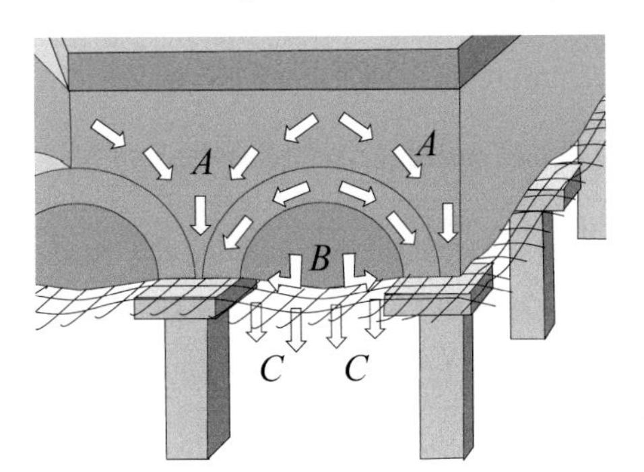

Fig. 2.1 Division of load into three parts (kN/pile): *A*: directly to the piles, *B*: via geosynthetic reinforcement to the piles, *C*: to the subsoil between the piles.

2.2 Deformations, settlements and differential settlements

Table 2.1 gives a summary of the current requirements on residual and differential settlements in the Netherlands. These requirements are based on a structure not founded on piles (traditional road foundation).

Table 2.1 Current requirements on residual settlement and differential settlement in the Netherlands, traditional road foundation.

Structure	Residual settlement	Differential settlement	Source
Motorway	0.15 to 0.30 m in 30 years	1:500, at transition zones 1:100	CROW 204
Secondary road	0.15 to 0.30 m in 30 years	1:70	CROW 204
Urban road	0.30 m in 20 years	1:70	CROW 204
Railway line	0.15 m in 10,000 days (max. 0.04 m in first year, 0.03 m in second year and 0.02 m in subsequent years)	1:333, track in ballast Stricter requirements apply to switches etc.	OVS 00056-7.1, version 3.0
Tramline	0.20 m in 20 years	1:500	Tramplus technical schedule of requirements (RET Public Transport Organisation, Rotterdam, Netherlands)
Airfield	0.03 m in 30 years	1:1500 over 45 m	CROW 204

The requirements in this table are relatively flexible for a structure founded on piles. The question is, would the client be better off with:

- stricter requirements in combination with a longer maintenance period, or
- the current requirements (Table 2.1), in combination with cheaper construction?

The deformation at the top of the structure (such as the road surface) comprises the sum of:

- the deformation of the pile head, and
- the deformation within the reinforced embankment and, if applicable, the road foundation.

The deformation within the reinforced embankment depends on the GR strain during the service life (after opening the road). For a good prediction of the service life deformations, two calculations are needed, the first for the end of the construction phase, and the second for the end of the service life, see Chapter 2.10.2.

2.3 Traffic load

2.3.1 Introduction

In the US, the typical traffic load used in design is 13 kPa while 20 kPa is commonly used in China. In the Netherlands, 15 kPa is often applied. These values are based on assumptions concerning the loaded road surface and the stress distribution in the soil structure. In relatively thin reinforced piled embankments however, a greater load can arise locally; while in higher piled embankments it can be reasonable to do the calculations for GR- and pile design with a lower traffic load.

This design guideline adopts the traffic load in load model BM1 in Eurocode NEN-EN 1991-2 [4] see Fig. 2.2. This BM1 load model has been converted, for various pile arrangements and embankment heights, into a uniformly-distributed load for the purposes of dimensioning the synthetic reinforcement in this type of embankment. Here, a distinction is made between situations involving one, or two or more traffic lanes. The results are shown in Table 2.2 and Table 2.3 and Appendix A. Chapter 2.3.2 describes how these tables should be used to determine the uniform design traffic load p (= $p_{traffic}$). Chapter 5.3 gives calculation examples.

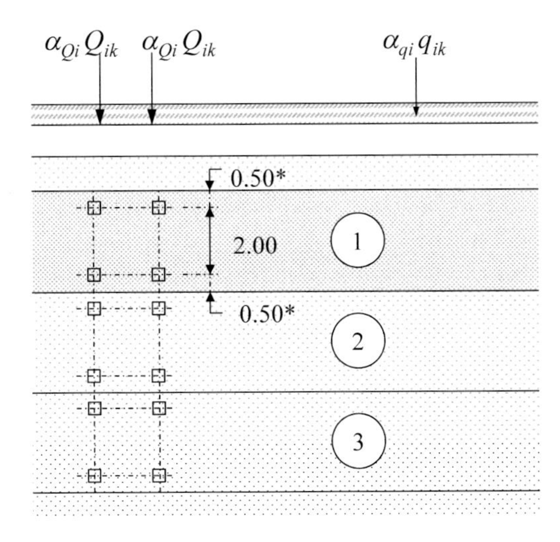

Explanation

(1) Traffic lane number 1: Q_{1k} = 300 kN; q_{1k} = 9 kPa

(2) Traffic lane number 2: Q_{2k} = 200 kN; q_{2k} = 2.5 kPa

(3) Traffic lane number 3: Q_{3k} = 100 kN; q_{3k} = 2.5 kPa; distance between axles in tandem pair = 1.2 m

(4) * for the theoretical value of the lane width: w_l = 3.0 m

Fig. 2.2 Traffic load: load model BM1 according to NEN-EN 1991-2+C1:2011 (Figure 4.2a – Application of load model 1)

According to NEN-EN 1991-2 the loads given in Fig. 2.2 are valid for the heaviest traffic intensity class. However, according to Art. 4.3.2 (3) of NEN-EN-1991-2, for the traffic compositions more usual on national trunk roads and motorways in the Netherlands, a reduction of up to 20% may be applied. For the traffic loads shown in Tables 2.3, 2.4 and those in Appendix A, this reduction of 20% is maintained. Tables 2.3, 2.4 and those in Appendix A should be adapted to the locally applicable traffic loads.

2.3.2 *Determination of uniformly distributed traffic load*

This Chapter presents the method for determining the characteristic value of the traffic load p.

1. Determine the traffic intensity (N according to NEN-EN 1991-2 [4]).
2. Determine the equivalent height H_{eq} over which the load may be distributed. This is determined from the height H (level of top surface/track bed to top of pile cap) and the relieving effect of the pavement construction layers (Braunstorfinger [14]), in accordance with the equations below (assuming two pavement construction layers; see also the example calculation in Chapter 5):

$$H_{eq} = h_{1;eq} + h_{2;eq} + h_3 \tag{2.1}$$

$$h_{1,2;eq} = 0.9 \cdot h_{1,2} \cdot \sqrt[3]{\frac{E_{1,2}}{E_3}} \tag{2.2}$$

where:

H_{eq}	the total equivalent layer thickness in m,
$h_{1,2;eq}$	the equivalent layer thicknesses for pavement construction layers 1 and 2 respectively in m,
h_3	the remaining embankment thickness, being H minus the thickness of the pavement construction layers in m,
$E_{1,2,3}$	the dynamic stiffness modulus of the various layer in MPa, where it applies that $E_3 = 200$ MPa.

3. Choose the square pile arrangement.
4. Determine the average maximum load $p_{traffic}$ (after distribution) on the geosynthetic reinforcement. For one lane, this value should be obtained from Table 2.2, and for two or more lanes the value should be obtained from Table 2.3. These tables apply for the traffic intensity N = 2,000,000. Appendix A provides tables for other traffic intensities. The tables are valid for a square pile arrangement. For rectangular pile arrangements, the

smallest spacing is normative (calculate with pile arrangement $s_x \cdot s_x$ if $s_x < s_y$), or the uniform traffic load may be derived from BM1 (with a reduction of 20% in the correction factor), in which the assumption may be made of a distribution according to Boussinesq over H_{eq} (as described in Van Eekelen *et al.*, 2010, [19]).

5. For axle loads greater or less than those in BM1, the applicable traffic load may be calculated by linear proportion.
6. In the ultimate limit state (ULS), the partial factors according to Chapter 2.7 should be applied to the uniform traffic load determined using the above relationships.

2.3.3 Traffic load summary

With (equivalent) heights greater than 1.0 m, the stress distribution proceeds more or less uniformly, and the load may be determined by following the steps above.
With (equivalent) heights less than 1.0 m, the stress distribution is so strongly dependent on the positions of the wheel loads that the above tables are no longer applicable. Consequently, the stress distribution needs to be determined for normative locations based on Boussinesq. The additional spreading capacity effect of the relatively stiff top layers may be included in this analysis by using a virtual extra height as shown in Chapter 2.3.2. The maximum load averaged over a pile unit with area A_i may also be assumed.

2.4 Horizontal load

Horizontal loads arising from braking forces should be taken into account at the end of a road structure, or in a bend. This also applies to centrifugal forces from the traffic that are the result of turning through an arc radius. In both cases, the traffic load determined using Chapter 2.3 should be increased by 20% for the GR design (not the pile design).

Horizontal load on the geosynthetic reinforcement and/or piles also arises from lateral thrust under the embankment slope; see Chapters 4.3.3 and 6.2.

Table 2.2 Maximum average uniform traffic load ptraffic on a basal reinforced piled embankment based on NEN-EN 1991-2 | N = 2,000,000 for lane 1, heavy traffic (F_{wheel} = 120 kN and $q_{uniform}$ = 7.2 kPa).

H_{eq} (m)	$p_{traffic;max}$ (kPa)	$p_{traffic}$ (kPa) for pile arrangement (m^2)					
		0.5 x 0.5	1.0 x 1.0	1.5 x 1.5	2.0 x 2.0	2.5 x 2.5	3.0 x 3.0
1.00	69.08	65.22	56.86	52.23	47.33	46.59	41.70
1.20	55.27	53.60	49.82	46.36	43.30	42.16	38.11
1.40	47.26	46.60	43.99	41.59	39.66	38.34	34.89
1.60	41.65	41.13	39.28	37.71	36.34	34.98	32.00
1.80	37.10	36.73	35.45	34.50	33.30	32.00	29.40
2.00	33.41	33.16	32.35	31.78	30.53	29.32	27.07
2.20	30.42	30.27	29.77	29.23	28.02	26.93	24.96
2.40	27.97	27.86	27.40	26.87	25.74	24.77	23.06
2.60	25.75	25.64	25.20	24.70	23.68	22.83	21.35
2.80	23.69	23.59	23.18	22.73	21.82	21.08	19.80
3.00	21.81	21.72	21.35	20.95	20.15	19.50	18.39
3.20	20.11	20.02	19.69	19.35	18.65	18.08	17.11
3.40	18.56	18.49	18.20	17.90	17.29	16.80	15.96
3.60	17.18	17.11	16.86	16.60	16.06	15.64	14.91
3.80	15.93	15.87	15.65	15.42	14.96	14.59	13.95
4.00	14.80	14.75	14.56	14.36	13.96	13.63	13.08
4.20	13.79	13.75	13.58	13.40	13.05	12.77	12.28
4.40	12.87	12.84	12.69	12.54	12.23	11.98	11.55
4.60	12.04	12.01	11.89	11.75	11.48	11.26	10.89
4.80	11.29	11.27	11.15	11.04	10.80	10.61	10.27
5.00	10.61	10.59	10.49	10.39	10.18	10.01	9.71
5.20	9.99	9.97	9.88	9.79	9.61	9.46	9.20
5.40	9.42	9.41	9.33	9.25	9.09	8.95	8.72
5.60	8.91	8.89	8.82	8.75	8.61	8.49	8.28
5.80	8.43	8.42	8.36	8.29	8.16	8.06	7.87
6.00	8.00	7.98	7.93	7.87	7.76	7.66	7.50
6.20	7.59	7.58	7.53	7.48	7.38	7.30	7.15
6.40	7.22	7.21	7.17	7.12	7.03	6.96	6.82
6.60	6.88	6.87	6.83	6.79	6.71	6.64	6.52
6.80	6.56	6.56	6.52	6.48	6.41	6.35	6.24
7.00	6.27	6.26	6.23	6.20	6.13	6.07	5.98
7.20	6.00	5.99	5.96	5.93	5.87	5.82	5.73

H_{eq} (m)	$p_{traffic;max}$ (kPa)	$p_{traffic}$ (kPa) for pile arrangement (m^2)					
		0.5 x 0.5	1.0 x 1.0	1.5 x 1.5	2.0 x 2.0	2.5 x 2.5	3.0 x 3.0
7.40	5.74	5.74	5.71	5.68	5.63	5.58	5.50
7.60	5.50	5.50	5.48	5.45	5.40	5.36	5.28
7.80	5.28	5.28	5.26	5.23	5.19	5.15	5.08
8.00	5.07	5.07	5.05	5.03	4.99	4.95	4.89

$p_{traffic,max}$ is the maximum vertical load on the GR;

$p_{traffic}$ is the maximum average load on the GR. This value is used as the characteristic value of the traffic load p and to determine the value of κ for the reduction of the arching. Possibly, a permanent load may be added to this, so that the design load p becomes: $p = p_{traffic} + p_{permanent}$.

Table 2.3 Maximum average uniform traffic load $p_{traffic}$ on a basal reinforced piled embankment based on NEN-EN 1991-2 | N = 2,000,000 for lane 1, heavy traffic (F_{wheel} = 120 kN and $q_{uniform}$ = 7.2 kPa) and lane 2 (F_{wheel} = 100 kN and $q_{uniform}$ = 2.5 kPa).

H_{eq} (m)	$p_{traffic;max}$ (kPa)	$p_{traffic}$ (kPa) for pile arrangement (m^2)					
		0.5 x 0.5	1.0 x 1.0	1.5 x 1.5	2.0 x 2.0	2.5 x 2.5	3.0 x 3.0
1.00	81.02	76.50	74.99	70.66	62.11	52.78	47.33
1.20	70.20	68.89	66.99	62.66	56.10	49.16	44.25
1.40	63.21	62.15	59.96	56.11	51.00	45.73	41.51
1.60	56.97	55.95	53.85	50.62	46.61	42.52	38.87
1.80	51.32	50.41	48.58	45.94	42.81	39.54	36.46
2.00	46.33	45.55	44.04	41.94	39.43	36.77	34.17
2.20	42.01	41.37	40.14	38.44	36.41	34.22	32.01
2.40	38.26	37.74	36.74	35.36	33.70	31.88	30.00
2.60	35.00	34.58	33.76	32.63	31.25	29.72	28.13
2.80	32.15	31.80	31.13	30.19	29.04	27.75	26.38
3.00	29.64	29.35	28.80	28.01	27.04	25.94	24.75
3.20	27.42	27.18	26.71	26.05	25.23	24.28	23.25
3.40	25.44	25.24	24.85	24.28	23.57	22.76	21.86
3.60	23.68	23.50	23.17	22.68	22.07	21.36	20.57
3.80	22.08	21.94	21.65	21.23	20.70	20.08	19.39
4.00	20.65	20.52	20.27	19.90	19.44	18.90	18.29
4.20	19.34	19.23	19.01	18.70	18.29	17.81	17.27
4.40	18.16	18.06	17.87	17.59	17.23	16.81	16.33
4.60	17.08	16.99	16.83	16.58	16.26	15.89	15.46
4.80	16.09	16.02	15.87	15.65	15.37	15.03	14.65

H_{eq} (m)	$p_{traffic;max}$ (kPa)	$p_{traffic}$ (kPa) for pile arrangement (m^2)					
		0.5 x 0.5	1.0 x 1.0	1.5 x 1.5	2.0 x 2.0	2.5 x 2.5	3.0 x 3.0
5.00	15.19	15.12	14.99	14.79	14.54	14.25	13.90
5.20	14.36	14.30	14.18	14.01	13.78	13.51	13.21
5.40	13.59	13.54	13.43	13.28	13.08	12.84	12.56
5.60	12.88	12.84	12.74	12.60	12.42	12.21	11.96
5.80	12.23	12.19	12.10	11.98	11.82	11.62	11.40
6.00	11.63	11.59	11.51	11.40	11.25	11.08	10.87
6.20	11.07	11.03	10.96	10.86	10.73	10.57	10.38
6.40	10.54	10.51	10.45	10.36	10.24	10.09	9.93
6.60	10.06	10.03	9.98	9.89	9.78	9.65	9.50
6.80	9.61	9.58	9.53	9.46	9.36	9.24	9.10
7.00	9.19	9.16	9.12	9.05	8.96	8.85	8.72
7.20	8.79	8.77	8.73	8.67	8.58	8.48	8.37
7.40	8.43	8.41	8.37	8.31	8.23	8.14	8.03
7.60	8.08	8.06	8.03	7.97	7.90	7.82	7.72
7.80	7.76	7.74	7.71	7.66	7.60	7.52	7.43
8.00	7.45	7.44	7.41	7.36	7.30	7.23	7.15

$p_{traffic,max}$ is the maximum vertical load on the GR;
$p_{traffic}$ is the maximum average load on the GR. This value is used as the characteristic value of the traffic load p and to determine the value of κ for the reduction of the arching. Possibly, a permanent load may be added to this, so that the design load p becomes: $p = p_{traffic} + p_{permanent}$.

2.5 Effects of dynamic loads

2.5.1 Dynamic loads on the piles

According to the current state-of-the-art, no account needs to be taken of the effects of dynamic loads on the piles in the design of basal reinforced piled embankments. Until now, this phenomenon has not led to any problems with basal reinforced piled embankments.

2.5.2 Dynamic loads on geosynthetic reinforcement

Dynamic loads on high-strength geosynthetic reinforcement made of PET, PVA or PP causes no change to the tensile strength of these materials. Therefore no additional reduction factor for the strength needs to be incorporated in the calculation of the long-term GR design strength.

2.5.3 *Effect of dynamic loads on arching*

If the variable load (traffic load) is no more than 50% of the total load, no account needs to be taken of the effect of the dynamic load on the arching; in equation form (design values):

$$\kappa = 1.0 \text{ for } \frac{p_{traffic}}{\sigma_{v;tot}} \leq 0.50 \quad \text{with} \quad \sigma_{v;tot} = \sum \gamma \cdot H + p \tag{2.3}$$

where:

$\sigma_{v;tot}$	total vertical stress at the level of the geosynthetic reinforcement neglecting the arching in kPa
γ	embankment unit weight in kN/m^3
H	embankment height in m,
$p_{traffic}$	the traffic load on the road surface (obtained from Table 2.2, Table 2.3 or the tables in Appendix A) in kPa,
p	the total permanent and variable surface load ($p_{traffic} + p_{permanent}$) in kPa.

For Dutch railways, where the uniformly distributed traffic load is prescribed, this means a minimum thickness of the reinforced embankment of 2.5 m (distance between bottom of track and bottom layer of reinforcement) is required to prevent the effect of the dynamic load having to be taken into account.

If the above condition is not met, then account should be taken of a reduction in arching in the calculations for the post-construction service condition. For this, use may be made of the κ - model of Heitz [16]. Due to the dynamic load, the load on the piles reduces (A becomes smaller) and the load on the subsoil and reinforcement increases (B+C becomes larger). It is indicated in Chapter 4.3.2.6 how this may be incorporated into the calculation with the help of the factor κ.

For the construction phase, no reduction due to dynamic effects needs to be taken into account due to the small number of traffic movements in this phase. For the construction phase $\kappa = 1.0$.

2.6 Safety approach

The safety approach is presented in the flowchart below (Fig. 2.3).

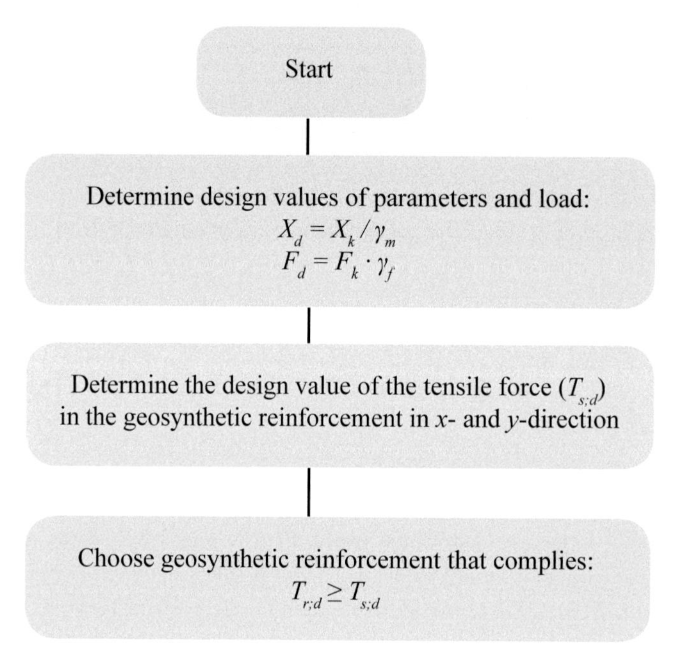

Fig. 2.3 Safety approach

The design calculations should be conducted for the embankment's service phase and one or more construction phases. In these, a distinction should be made between the ultimate limit state (ULS) and the serviceability limit state (SLS).

Reliability classes

Table 2.4 presents an overview of the reliability classes. In the Netherlands, a basal reinforced piled embankment for roads is usually categorised into reliability class RC1. Railways normally demand RC3. This depends on the envisaged time span within which the structure should continue to meet the requirements (reference period) and the possible ramifications of failure.

Table 2.4 Reliability classes, Eurocode / NEN.

Eurocode Reliability Class with corresponding reliability index β	Service function (see NEN-EN 1990/NB and others)	Design service life (years)
RC1 ($\beta = 3.3$)	Agricultural buildings, horticulture greenhouses, standard single-family homes, industrial buildings (one or two storeys)	15
	Roads	50
RC2 ($\beta = 3.8$)	Residential, office, public and industrial buildings (three or more storeys)	50 to 100
RC3 ($\beta = 4.3$)	High-rise buildings ($h > 70$ m), stadiums, exhibition spaces, concert halls, large public buildings, railways	50 to 100

Normative construction phases

The calculations should be conducted for all possible phases, including the construction phase. During the construction phase, the effects of traffic load (construction traffic) may be great while the embankment is still relatively low. The geosynthetic reinforcement will still react relatively strongly and stiffly in the construction phase given that little strain has yet arisen due to creep.

A different Reliability Class may apply during the construction phase from that during the service phase. This should be assessed in each case, with the risk of failure being considered. In this, a distinction is made between personal safety risks and economic risks.

Failure in the construction phase could for example mean that the geosynthetic reinforcement fails locally due to an excessive load. Both the personal safety risk and the economic risk are usually slight in this phase. This may be a reason to opt for a lower Reliability Class for the construction phase. However, this choice cannot be declared to be generally applicable.

As stated earlier, the factor κ, to take into account dynamic effects on arching, may be set to 1.0 in the construction phase (see 2.5.3).

2.7 Partial factors

The partial factors needed for the calculation of the ultimate limit state (ULS) are presented in the tables below. In the serviceability limit state (SLS), all the load and material factors are equal to 1.0.

When applying the Concentric Arches calculation model as discussed in Chapter 4.3, a model factor shall be applied. The model factor is defined based on correlations of values calculated with the calculation model and values measured in experimental laboratory setups and piled embankments realised in practice and a Monte Carlo analysis.

The partial factors used for the calculation of the geosynthetic reinforcement have also been derived based on Monte Carlo analyses. For the derivation of both the model and the partial factors, reference is made to Van Duijnen *et al.*, 2015 [18]. Table 2.5 presents the values found for the model and partial factors.

The load factors for the design of the piles are listed in Table 3.1.

Table 2.5 Load and material factors for GR dimensioning.

Parameter	Factor	SLS	Reliability class in ULS		
			RC1	RC2	RC3
		($\beta \geq 2.8$)	($\beta \geq 3.5$)	($\beta \geq 4.0$)	($\beta \geq 4.6$)
Model factor	γ_M	1.40	1.40	1.40	1.40
Traffic load, p	$\gamma_{f;p}$	1.00	1.05	1.10	1.20
Tangent of internal friction angle, tan φ'	$\gamma_{m;\varphi}$	1.00	1.05	1.10	1.15
Unit weight of fill material, γ *	$\gamma_{m;\gamma}$	1.00	0.95	0.90	0.85
Subgrade reaction of subsoil, k_s	$\gamma_{m;k}$	1.00	1.30	1.30	1.30
Axial tensile stiffness of geosynthetic reinforcement, J	$\gamma_{m;J}$	1.00	1.00	1.00	1.00
Tensile strength of geosynthetic reinforcement, T_r	$\gamma_{m;T}$	1.00	1.30	1.35	1.45

γ_M is the model factor by which the calculated GR strain and GR tensile force should be multiplied,

γ_f is the load factor, $F_d = \gamma_f \cdot F_k$,

γ_m is the material factor, $X_d = X_k / \gamma_m$,

* Increasing the unit weight is not beneficial, hence the value of $\gamma_{m;\gamma}$ is less than 1.0.

2.8 Pile location tolerance

In determining the appropriate partial factors from Table 2.5, account is taken of a location tolerance of ± 0.10 m in the horizontal plane and ± 0.05 m in the vertical plane. In the calculation of the axial pile load, this tolerance does not need to be taken into consideration.

2.9 Service life / maintenance

The service life of the structure depends on the service life of the various components. Normative here are the geosynthetic reinforcement and the reinforcement in the piles.

Suppliers of geosynthetic reinforcement indicate that a service life of at least 120 years may be expected. The supplier of the entire system (reinforced embankment and piles) will require justifying the service life, taking account of the specific environmental conditions; see Chapters 2.10.1 and 4.2.3 (GR reduction factors). For the reinforcement of precast concrete and cast-in-situ piles, use may be made of the existing standards.

The technical service life of the structure is often longer than the economic service life.

In a basal reinforced piled embankment, only the asphalt or the railway track ballast and track will require periodic maintenance. In most cases, more maintenance will be needed where there are transition zones to structures that undergo additional settlement.

2.10 Checking the geosynthetic reinforcement (GR)

2.10.1 Tensile strength

To determine the design values of the GR tensile strengths to be used, there are two methods available, the British Standard BS 8006 [10] and the German EBGEO [11], see Table 2.6. Both design codes employ the same principles in determining the long-term GR tensile strengths.

Table 2.6 Determination of characteristic long-term tensile strength of geosynthetic reinforcement according to EBGEO and BS 8006.

EBGEO [11]	BS 8006 [10]
$T_{r;lt;k} = \frac{T_{r;st;k}}{A_1 \cdot A_2 \cdot A_3 \cdot A_4 \cdot A_5}$	$T_{r;lt;k} = \frac{T_{r;st;k} \cdot RF_{creep}}{f_{m11} \cdot f_{m12} \cdot f_{m21} \cdot f_{m22}}$

where:

$T_{r;st;k}$ is the short-term characteristic value of the GR tensile strength in kN/m. This is the short-term strength of the geosynthetic reinforcement at the end of production at the factory, with a certainty of 95% to 99%. The certainty margin may vary for different suppliers,

$T_{r;lt;k}$ is the long-term characteristic value of the tensile strength of the geosynthetic reinforcement taking into account reduction factors for mechanical damage and environmental effects, in kN/m,

A_1 is the reduction factor for the load duration (creep),

A_2 is the reduction factor for damage during transport, installation and compaction,

A_3 is the reduction factor for connections and welded seams. $A_3 = 1.0$ when the force transfer proceeds via a continuous layer of reinforcement. When two reinforcement layers are connected together, the connection should be tested. If it emerges that the strength of the connection is only 50% of the original strength of the geosynthetic reinforcement for example, then A_3 becomes 2.0 for the connection-area to take a reduced strength into account. A connection with another element also requires testing in advance, and the new A_3 for this connection then follows,

A_4 is the reduction factor for the influence of the environment,

A_5 is the reduction factor for the variable load (usually $A_5 = 1.0$, see Chapter 2.5.2),

RF_{creep} is the reduction factor for the duration the load is applied (creep),

f_{m11} is the reduction factor for possible errors in the mathematical model,

f_{m12} is the extrapolation factor based on the period of test results and takes account of the confidence of the long-term capacity assessment. This factor may vary with the required service life of the reinforcement,

f_{m21} is the reduction factor for damage during transport, installation and compaction,

f_{m22} is the reduction factor for the influence of the environment.

The reduction factors A_1 to A_5, RF_{creep} and f_{m11} to f_{m22} should be determined for the specific reinforcement material and is usually provided by its supplier. It should be demonstrated that these reduction factors have been determined in accordance with the applicable standards and guidelines.

For the construction phase, values of A_1, A_4, RF_{creep} and f_{m22} are different from those over the service life. This has the consequence that in the construction phase it is usually possible to calculate with a higher tensile strength than that over the service life phase.

After establishing the characteristic value of the long-term tensile strength of the geosynthetic reinforcement, the design value of the long-term characteristic tensile strength may be determined by application of the partial material factor $\gamma_{m;T}$, given in Chapter 2.7:

$$T_{r;lt;d} = T_{r;lt;k} / \gamma_{m;T} \tag{2.4}$$

where:

$T_{r;lt;d}$ is the factored design value of the long-term characteristic tensile strength of the geosynthetic reinforcement in kN/m,

$T_{r;lt;k}$ is the characteristic value of the long-term tensile strength according to Table 2.6 in kN/m,

$\gamma_{m;T}$ is the partial material factor for the strength of the geosynthetic reinforcement according to Table 2.5.

Checking the tensile strength of the geosynthetic reinforcement
The design tensile strength of the geosynthetic reinforcement should be equal to or greater than the tensile force applied in both directions:

$$T_{r;lt;d} \geq T_{s;d} \tag{2.5}$$

where:

$T_{r;lt;d}$ is the long-term design value of the tensile strength of the geosynthetic reinforcement in kN/m,

$T_{s;d}$ is the design value of the total tensile force in the geosynthetic reinforcement in the x or y direction as appropriate (see chapter 4.3.4) in kN/m.

2.10.2 *Deformation*

The average strain in the geosynthetic reinforcement for the SLS (serviceability limit state) at the end of the service life of the structure, calculated according to Chapter 4.3, should be less than the maximum permissible strain of the reinforcement.

The strain during the in-service phase should be less than the maximum permissible strain during the in-service phase, and this strain is calculated as the difference between:

- The average SLS strain immediately after construction, before opening the road:
  - no traffic load: $p = 0$ kPa;
  - possible subsoil support;
  - GR stiffness according to the isochronous curves, including time-dependence.
- The average SLS strain at the end of the service life:
  - the traffic load p determined according to the client's specifications and Chapter 2.3;
  - less or no subsoil support;
  - GR stiffness according to the isochronous curves, including time-dependence.

This is shown in the calculation example of Chapter 5.

2.11 *Subsoil support of the geosynthetic reinforcement*

In some cases, the reinforced embankment is permanently supported by the subsoil between the piles. This support may be included by including subgrade reaction k_s, as described in Chapter 4.3. In this case, the piles are nonetheless designed for all loads (see Chapter 3.2). In many cases, permanent subsoil support cannot be guaranteed at the end of the service life (i.e.: $k_s = 0$ kN/m^3) due to subsidence of the subsoil over time. For example, the weight of the working platform beneath the GR may induce a gap to form under the GR.

2.12 *Surface water*

If a piled embankment is used for a railway line, surface water management is an important design detail. In contrast to road structures, surface water can percolate through the entire upper surface.

When using crushed demolition waste aggregate which can have low hydraulic conductivity surface water should be allowed to flow away at the sides. The sides should therefore not consist of low hydraulic conductivity layers (e.g. peat or clay).

If the bottom of the reinforced embankment lies at a deep level (at or near groundwater level) and the sides consist of peat or clay, the risk exists that water will remain standing below the track. In this situation, the track's stability is not guaranteed.

2.13 *Applicability and limitations of this design guideline*

This design guideline is intended for basal reinforced piled embankments with geosynthetic reinforcement. The design rules described in this design guideline have been validated with measurements conducted in a number of practical projects and experimental series ([22] Van Eekelen *et al.*, 2015, also included in Van Eekelen, 2015 [23]). The model factor and partial material and load factors belong to the calculation rules given in this design guideline and were determined based on a series of probabilistic analyses ([18] Van Duijnen *et al.*, 2015).

Table 4.2 presents a number of boundary conditions that the piled embankment should meet if this design guideline is to apply. The following should also be noted: the validation of the design rules for the geosynthetic reinforcement was conducted with measurements on piled embankments:

- with a centre-to-centre spacing ≤ 2.50 m;
- where reinforcement layers of geogrids, possibly combined with woven geotextile, were applied (geogrid on geotextile);
- where the groundwater level was below or only slightly above the pile caps;
- where $0.5 \leq H/(s_d - d_{eq}) \leq 4.0$));
- with vertical stresses on the pile cap of up to 1450 kPa. In practice, however, some embankments of this type have already been realised with vertical stresses on the pile cap of 2000 kPa.

The validated application of the design rules described in this guideline is limited to embankments whose geometry and materials used meet the above boundary conditions. For non-compliant geometries or combinations of materials, additional measurements and/or suitability tests should be or should have been conducted in representative practical cases with which it can be demonstrated that the system comes within the framework of this design guideline.

3 Requirements and initial details for the piles and pile caps

3.1 Introduction

Piles and pile caps should be designed according to locally applicable design rules. For this reason, this chapter confines itself to the requirements and initial details for the design of the piles and pile caps.

3.2 Piles

The piles should be designed following local guidelines. In the transition zone to non-piled road sections, the pile spacing is sometimes increased and/or the pile toe depth is reduced gradually. Any loads arising from negative skin friction of subsiding soil layers should be taken into account according to locally applicable guidelines.

The piles should be designed such that they can bear the entire load: traffic load (calculated according to Chapter 2.3), weight of the road pavement structure and embankment weight.

Table 3.1. lists the load factors for the design of the piles in accordance with the Eurocode procedure. In the design calculations for the piles, all loads are carried by the piles. The piles are further designed to meet the locally applicable rules.

Table 3.1 Load factors to calculate the bearing capacity of the pile foundation.

EuroCode Reliability Class	**No permanent subsoil support beneath the reinforced embankment**		**Permanent subsoil support beneath the reinforced embankment***	
	$\gamma_{f;g}$	$\gamma_{f;q}$	$\gamma_{f;g}$	$\gamma_{f;q}$
RC1 ($\beta = 3.3$)	1.1	1.35	1.0	1.0
RC2 ($\beta = 3.8$)	1.2	1.5	1.1	1.1
RC3 ($\beta = 4.3$)	1.3	1.65	1.2	1.2

$\gamma_{f;g}$ is the load factor for permanent load, $F_{g;d} = \gamma_{f;g} \cdot F_{g;k}$

$\gamma_{f;q}$ is the load factor for variable load factor, $F_{q;d} = \gamma_{f;q} \cdot F_{q;k}$

* with the assumption that all load ends up on the piles

To determine the bearing capacity of a pile, it is important whether the reinforced piled embankment may be considered as a stiff structure or not. Most local design guidelines for this reason feature a partial factor for the pile design that takes into account whether the structure is stiff or non-stiff. A reinforced piled embankment may be considered as a stiff structure if, when one pile fails, it is still possible to meet boundary condition 1 of Table 4.2, and detailed in Table 3.2. Thus, if one pile fails, the arching will re-establish itself. If the embankment is sufficiently high, the load will be transferred to the surrounding piles via this newly re-established arching. In this case, the normative centre-to-centre distance between the pile caps has increased by a factor of 2. This is illustrated in Fig. 3.1.

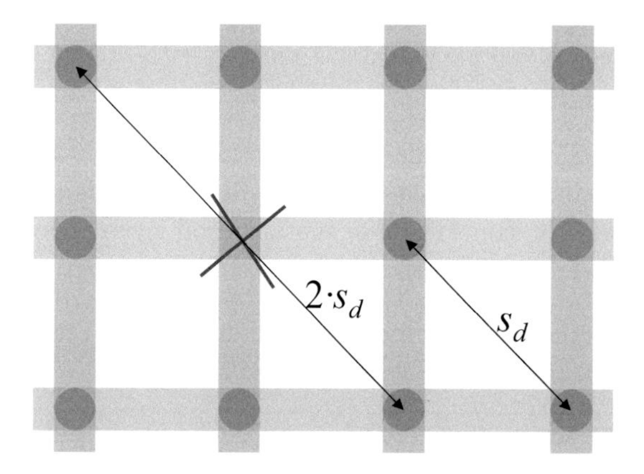

Fig. 3.1 Increase in diagonal centre-to-centre spacing of the piles s_d if one pile fails.

Table 3.2 Initial details for the bearing capacity calculation in pile design for a basal reinforced piled embankment.

$H/(2\cdot s_d - d_{eq}) < 0.66$	non-stiff piled embankment: reinforced embankment should be considered as a flexible superstructure in pile bearing capacity calculations
$H/(2\cdot s_d - d_{eq}) \geq 0.66$	stiff piled embankment: reinforced embankment can be considered as a rigid superstructure in pile bearing capacity calculations

When the reinforced embankment is non-stiff, which will often be the case for shallow embankments where the pile spacing is maximised, the design should be done using the factors appropriate for a non-stiff structure, unless it has been demonstrated that the structure will continue to perform if one pile fails. It should be noted that the consequences of a pile failure will in most cases result in larger deformations and not result in the collapse of the structure overall.

Deformations, forces and bending moments in the piles can be determined using a numerical model, such as the FEM model; see Chapter 6.

3.3 Pile caps

A pile cap may be a separate pre-cast element placed on top of the pile beneath the reinforced embankment, or it may be an integral part of the foundation pile, for example cast-in-situ piles - see Fig.1.2 and Chapter 8.2.

The dimensions of the pile cap and the way in which the connection with the pile is made are important for the load distribution in the reinforced embankment and the pile. The detailing of the connection partly determines the bending moments that are transferred into the piles.

The pile cap should be dimensioned for:

- bending moment;
- horizontal normal force;
- vertical transverse force or punching.

The various loads acting on the pile caps are shown in Fig. 3.2.

Also, it should be ensured that the edge of the pile caps cause negligible damage to the geosynthetic reinforcement. This can be achieved by rounding off the edges of the pile caps and/or providing protection between the geosynthetic reinforcement and the pile caps.

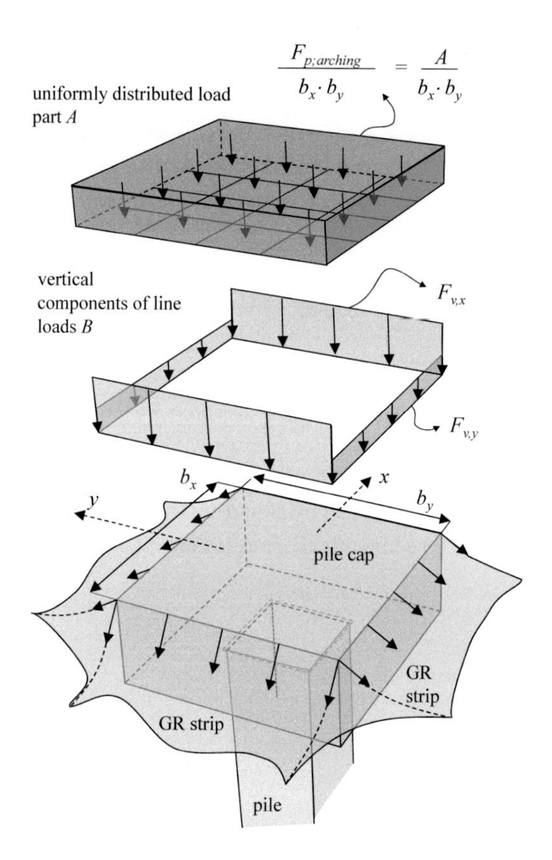

Fig. 3.2 Load on pile caps.

4 Design of the reinforced embankment

4.1 Introduction

The basal reinforced embankment that is constructed on top of the pile group consists of an embankment that is reinforced at its base with one or more geosynthetic reinforcement layers; see Fig. 4.1. The bottom fill layer of the embankment (of height h^* - see Table 4.2) incorporating the reinforcement is also called a 'mattress' or a 'load transfer platform'. Chapter 4.2 of this chapter presents the requirements for the reinforced embankment. Chapter 4.3 describes how the geosynthetic reinforcement should be designed.

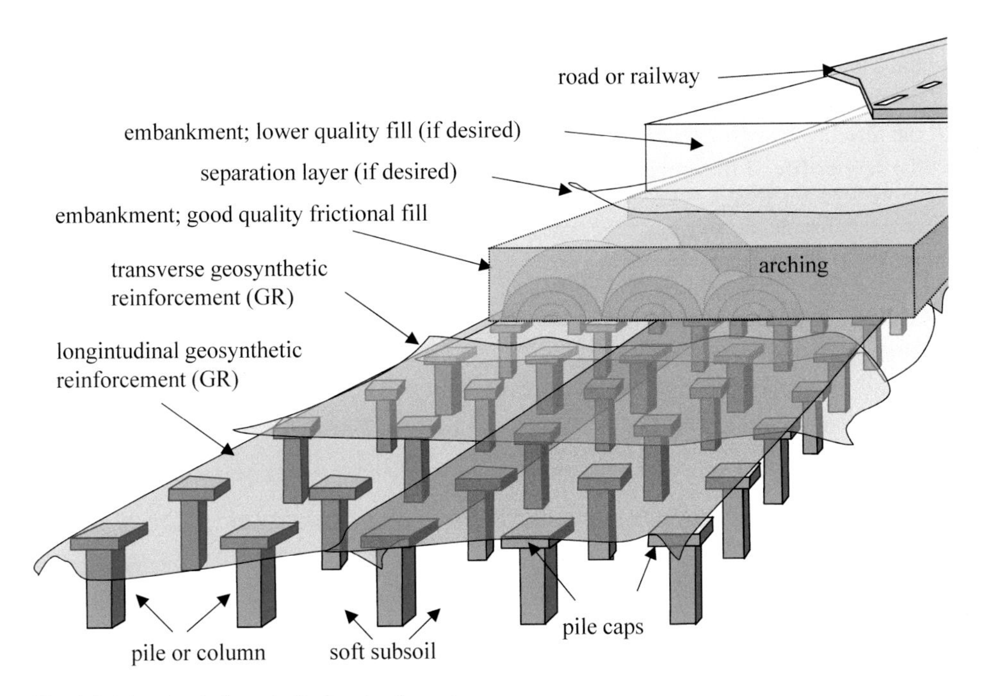

Fig. 4.1 A basal reinforced piled embankment

Very early in the design process, an optimisation of the pile group layout in relation to pile costs and the costs of the reinforced embankment should be undertaken.

4.2 Requirements for the basal reinforced embankment

4.2.1 Introduction

The primary function of the reinforced embankment is to transfer the greater part of the vertical load from the pavement or rail construction (and any traffic loads) to the piles. The embankment should fulfil this function over its entire service life without unacceptable deformations occurring. The embankment may only be able to achieve this if the granular fill remains stable and the top fill layers cannot be eroded (see Fig. 1.1 and Fig. 4.1). Embankment requirements that need to be established are:

- the embankment height;
- the strength of the reinforcement and the shear resistance of the fill, including the bond length of the reinforcement;
- the reinforcement tensile stiffness and creep behaviour to ensure the reinforcement strain remains within acceptable limits;
- the chemical and mechanical interaction between reinforcement and granular fill;
- the drainage capability of the fill;
- the filtering properties of the embankment;
- the service life of the embankment.

Additional requirements may apply to the edges of the embankment in connection with its stability.

Besides these, additional requirements are imposed in connection with the construction phase, such as:

- the fill installation method and its compressibility;
- the workability of the reinforcement.

4.2.2 Strength and stiffness of geosynthetic reinforcement

Chapter 4.3 describes how the necessary GR strength and GR stiffness should be determined.

To restrict the deformation of the embankment while in service, constraints are often imposed on the strain that the reinforcement may undergo (SLS), such as:

- the maximum total GR strain, for example $\varepsilon \leq 6\%$ (SLS). This requirement is prescribed in BS 8006 [10]. For shallow embankments, a stricter requirement may be necessary to prevent uneven deformations at the surface. In practice, a maximum strain of 3 to 4% is often used;
- the GR strain during the service life (SLS) as a result of traffic load and creep, for example $\varepsilon \leq 2\%$. See Chapter 2.10.2.

The design value of the strength of the geosynthetic reinforcement should be greater than the design value of the GR tensile force (ULS).

- Chapter 2.10.1 describes the determination of the design value of the strength of the geosynthetic reinforcement; see also Chapters 4.2.3 and 4.3.2.3.
- The tensile force is the result of the vertical load (embankment weight plus traffic) and the horizontal load. The tensile force due to the vertical load is described in Chapter 4.3.2. The horizontal load may be the result of braking or centrifugal forces, for which see Chapter 2.4, as well as the lateral outward thrust in the slope of the embankment, see Chapter 4.3.3. The necessary reinforcement strength to resist the lateral outward thrust depends on the height of the embankment. In wide embankments, no additional strength is needed in the centre of the transverse direction.

Chapter 8 presents instructions for overlaps and anchorage lengths for the geosynthetic reinforcement.

In practice, different types of geosynthetics are used in reinforced embankments. The most widespread are geogrids (with large open spaces between the tensile strands) and woven geotextiles, see also Chapter 2.13.

4.2.3 Service life of the geosynthetic reinforcement

The service life of the embankment is primarily governed by the service life of the geosynthetic reinforcement. The granular fill is assumed to have unlimited service life.

For the geosynthetic reinforcement, account should be taken of creep, which causes the rupture strength to decrease and the strain to increase as the load duration increases. These two time aspects (creep deformation and reduction in rupture strength) should be taken into account in the design. A safe design should ensure that at the end of the service period the reinforcement still has sufficient remaining strength and the creep strain is controlled. The properties of the geosynthetic reinforcement may also be adversely affected by certain environmental conditions, installation damage and temperature. Once these circumstances are recognised, account can be taken of them in the design. The manufacturer of the geosynthetic reinforcement can provide the necessary material reduction factors (for example A_1, A_2, A_3, A_4 and A_5) to account for different conditions. Chapter 2.10.1 describes the determination of the design value of the tensile strength of the geosynthetic reinforcement by making use of these reduction factors.

4.2.4 *Overlaps of the geosynthetic reinforcement*

A roll of geosynthetic reinforcement material is usually at most 5 m wide and 100 to 300 m long.

Overlaps should be avoided in the longitudinal direction (the strong direction). If an overlap in the strong direction cannot be avoided, an overlap of at least three pile rows is recommended, in other words two pile spans; see Chapter 8.

Where there are transitions in the transverse direction (perpendicular to the strong direction), an overlap is also made. This overlap should lie above the pile caps and not in the span between them. Chapter 8 goes further into the requirements of overlaps.

Starting with the geosynthetic reinforcement, the optimal pile arrangement follows from:

$$\frac{\text{roll width (usually 5 m) - overlap}}{(n)\ \text{piles - 1}}$$

If the pile spacing chosen deviates from this dimension, the reinforcement material may be applied wastefully.

The amount of overlap may be determined using the relationship below:

$$\sigma_v' \cdot \mu \cdot L \geq T_{TD} \tag{4.1}$$

where:

σ_v'	kPa	average vertical effective pressure on the pile cap,
μ	-	friction factor,
L	m	amount of overlap ($L \geq 0.2$ m),
T_{TD}	kN/m	the tensile strength of the reinforcement in the transverse direction of the roll.

The construction tolerance should be added to the design value of L. The friction factors (μ) between the adjacent reinforcement material, and between the geosynthetic and the pile cap and the aggregate are generally difficult to establish. When doubt exists about the friction factor to be used, use of equation (4.1) is not recommended. In this case the recommendations in Chapter 8 may be used.

4.2.5 *Embankment fill*

The embankment fill should have properties that enable it to be placed and compacted properly. In practice, this means that it should consist of granular material.

For good load distribution in the reinforced embankment, the fill should possess appropriate frictional properties (a large internal friction angle φ').

For a robust structure, whose arching is ensured under static and dynamic loads, coarse-grained material such as aggregate is suitable. The fill material should be able to resist the peak loads that may arise on top of the pile caps. It should be borne in mind that fine-grained material may be prone to erosion and its interaction with the geosynthetic reinforcement may break down under dynamic loads.

Proper interaction between the geosynthetic reinforcement and fill imposes specific requirements on both these components. A verified combination is coarse granular material (for example aggregate) with geogrids. The strength and stiffness of this material increases via the compaction of the subsequent layers.

The fill may display hydraulic binding, but this should not lead to a brittle, monolithic mass of low strength. A certain plastic deformation is necessary for proper load distribution throughout the embankment. Shrinkage should not lead to crack formation in the embankment either.

Internal friction angle of fill
Finite element calculations have shown that when a void forms under the geosynthetic, strains can arise in the fill that are so large (> 1.0 to 1.5%) that 'post-peak' behaviour can be expected. It is therefore unsafe to design using the peak value of the internal friction angle.

Conducting strength tests (triaxial tests) on crushed demolition waste aggregate is difficult and therefore few references are available. Two studies conducted in Delft, the Netherlands, involving large-scale vacuum triaxial equipment were analysed for this CUR guideline (Van Niekerk *et al.* [17] and Den Boogert (2011) [15]). In these, peak values (φ'_p) were measured, but the tests were not continued to a sufficiently high strain level to allow the constant volume shearing resistance (φ'_{cv}) to be determined. The tests reveal secant values of φ'_p ranging from 49 to 65 degrees (on separate Mohr circles) for cell confining stresses varying between 16 and 90 kPa. Based on the relationships for quartz sand (Bolton *et al.* [13]), a maximum difference between the peak value (φ'_p) and the constant volume value (φ'_{cv}) may be approximately 5 to 8 degrees, the exact values depending on the relative density and the isotropic stress. Assuming that this relationship is also valid for coarser, crushed recycled aggregate, a lower limit for φ'_{cv} of approximately 45 degrees may be assumed.

If sand is properly compacted, a characteristic value φ'_k of at most 35 degrees may be assumed. Thus, in a design where the majority of the arching occurs in sand fill, it is recommended to check the internal friction angle of the sand fill at various densities by means of laboratory tests.

Prescriptions and guidelines
The granular fill should consist of coarse-grained aggregate of good quality. If recycled aggregate is used, the recommended particle size grading is 0 / 31.5 or 4 / 31.5 mm to 0 / 63 or 4 / 63 mm, according to EN13242.

The following specific requirements are imposed on the fill material in proximity with the geosynthetic reinforcement:

- the fill should not contain any sharp objects that could damage the geosynthetic reinforcement;
- requirements on the service life, see Chapter 4.2.3;
- requirements from environmental legislation;
- requirements to do with the construction, see Chapter 8;
- requirements on compaction.

Concerning the design of the reinforcement, the following values for the residual internal friction angle (not peak values) of the fill material should be kept to (providing the requirements above are met):

- for sand, $\varphi'_k \leq 35$ degrees;
- for crushed demolition waste aggregate, $\varphi'_k \leq 45$ degrees.

It is only permitted to deviate from the above prescribed values when it is possible to demonstrate, by means of appropriate verification procedures in conformity with relevant standards, that higher values for the internal friction angle are appropriate.

In a layered construction, a weighted average of the φ' value may be required, for example according to NEN 9997-1 (foundations on steel), where the lowest layer of the piled embankment is the most important.

4.3 Analytical design of the basal geosynthetic reinforcement in the embankment

4.3.1 Introduction

The design procedure for a basal reinforced embankment is described in Table 1.1. The GR tensile force and the GR strain are determined in this Chapter using analytical relationships.

- The GR tensile force and GR strain as a result of the vertical load are described in Chapter 4.3.2.
- The GR tensile force as a result of the lateral outward embankment thrust is described in Chapter 4.3.3.
- Consequences of horizontal load that are the result of braking and centrifugal forces are covered in Chapter 2.4.
- Chapter 4.3.4 covers the total GR tensile force.

A numerical calculation method, such as the finite element method (FEM, e.g. Plaxis) may be used to calculate the bending moments in the piles and the deformations in the embankment. This is described in Chapter 6.

As described in Chapter 2.1, in determining the distribution of the vertical load over the piles, the geosynthetic reinforcement and the subsoil, distinction is made between the following (see Fig.2.1):

- load part A (kN/pile) that goes directly to the piles;
- load part B (kN/pile) that goes to the piles via the geosynthetic reinforcement;
- load part C (kN/pile) that is carried by the subsoil.

The embankment and the piles should be designed for all construction phases and the service life.

If multiple layers of geosynthetic reinforcement are used, then the strength and stiffness of the layers in the same direction may be summed up. See also Chapter 4.3.2.3.

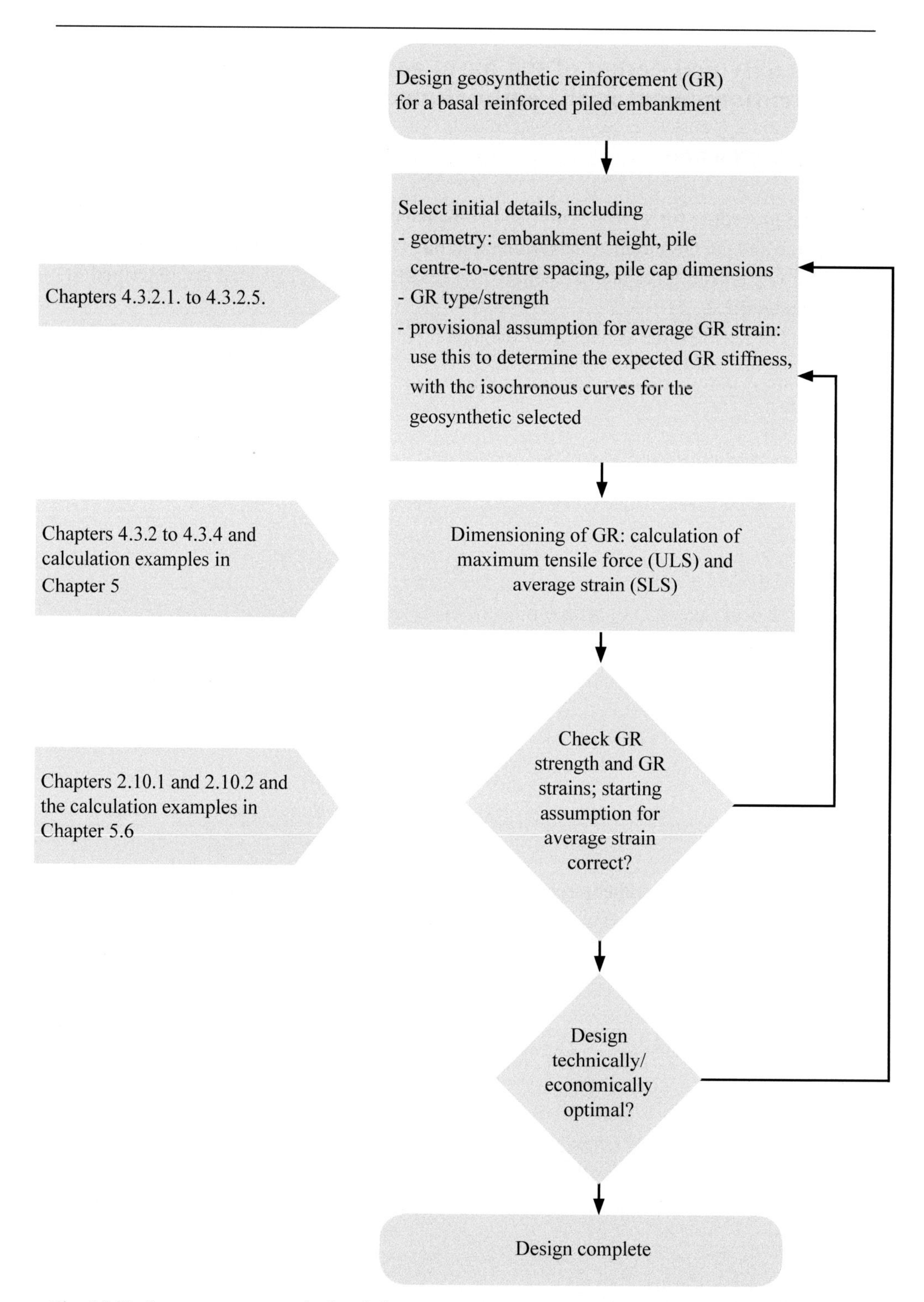

Fig. 4.2 Design process geosynthetic reinforcement (GR) for a basal reinforced piled embankment.

Fig. 4.2 presents the design process for the design analysis for the GR tensile strength and GR stiffness. The design process described in this guideline is only valid for a rectangular pile arrangement. An optimal design is obtained by conducting the design calculations in the following Chapters iteratively for all the required load cases.

4.3.2 GR tensile force and GR strain due to vertical load

This Chapter describes how the GR tensile forces and GR strains are to be determined Chapter 5 presents calculation examples.

The design process described in this document is largely adopted from Van Eekelen *et al.* (2012b, 2013 and 2015, [20], [21] and [22]) and Van Eekelen (2015, [23]) and partly from Chapter 9 of the German design guideline EBGEO [11].

4.3.2.1 Geometry

The dimensions described below should be determined (see Fig. 4.3 and Table 4.2).

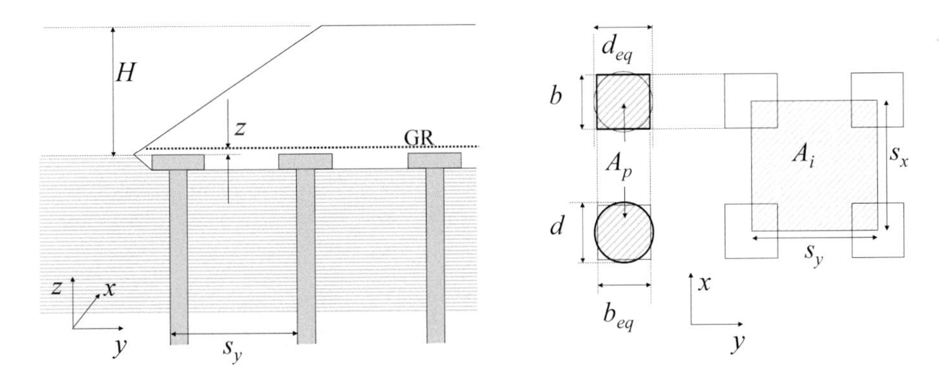

Fig. 4.3 Cross-section and plan view of piled embankment geometry

H	m	the height of the piled embankment in m, measured from the top of the pile cap to the top of the road surface,
s_x	m	the centre-to-centre spacing of the piles in the longitudinal direction of the embankment,
s_y	m	the centre-to-centre spacing of the piles in the transverse direction of the embankment,
A_i	m^2	area of influence of each pile, to be calculated according to

$$A_i = s_x \cdot s_y \qquad 4.2$$

s_d m the centre-to-centre spacing of the piles, equal to the diagonal of the area of influence A_i:

$$s_d = \sqrt{s_x^2 + s_y^2} \tag{4.3}$$

A_p m² area of the pile cap,

d m the diameter of the pile cap; if the pile cap is not round, d is made equal to the equivalent diameter d_{eq} according to:

$$d = d_{eq} = \sqrt{4 \cdot A_p / \pi} \tag{4.4}$$

b m the length of the side of the square pile cap; if the pile cap is round, b is made equal to the equivalent side b_{eq} according to:

$$b = b_{eq} = \frac{1}{2} \cdot d \cdot \sqrt{\pi} \tag{4.5}$$

z m the distance between the top of the pile cap and the geosynthetic reinforcement. When two layers of reinforcement are used, the average distance between each of the two layers should be used as the basis.

4.3.2.2 Subsoil parameters

The GR tensile forces will reduce if the subsoil bears part of the load. The subsoil's contribution depends on the subsoil stiffness which may be characterised by the subgrade reaction k_s (kN/m³). When this contribution can be guaranteed over the structure's required service life it may be included in the design analysis. If the subsoil's contribution cannot be guaranteed over the required service life a subgrade reaction of $k_s = 0$ kN/m³ should be used.

A different value of k_s generally applies during the construction phase from the value applied in service. It is necessary to include this difference to allow calculation of the difference in strain between that at the handover of the road and that at the end of its service life. This is described in Chapter 2.10.2.

4.3.2.3 Material properties of geosynthetic reinforcement

For the geosynthetic reinforcement, the maximum short-term tensile strength and the relationship between the GR tensile force and the associated GR strain should be known over the envisaged service life of the structure. The tensile stiffness and the tensile rupture strength are time-dependent due to creep effects. The relationship between tensile rupture strength, strain and their time-dependence are provided by the manufacturers of the geosynthetic reinforcement in the form of stress/strain isochronous curves. Standards and guidelines have been established for the preparation and presentation of these curves.

Using isochronous curves, the GR tensile stiffness can be determined for any load duration and at any strain level for the geosynthetic reinforcement. This tensile stiffness should be determined for the two main orientations, parallel and perpendicular to the road axis. For clarification, a calculation example is shown in Table 4.1. This calculation example assumes two situations, the construction phase (estimated duration seven days) and the service life phase. In the example, the required service life of the structure is taken to be 120 years. The example estimates a maximum strain of 4% for the deformation requirements that are imposed on the strain, and serve here only as an example entry.

Usually, reinforcement is used that has a notably higher strength in one direction (the machine direction) than in the other. Because sufficient strength in both directions should be present in the mattress, two layers of reinforcement are commonly used, one with its high strength in the road's transverse direction and the other in the road's longitudinal direction. It is permitted to design with the total strength in a particular direction, in other words:

- total strength in road's transverse direction = sum of high and low strengths in transverse direction;
- total strength in road's longitudinal direction = sum of high and low strengths in longitudinal direction.

If any material is connected or seamed together, the above approach should only be applied if due account is taken of the seam or connection strengths attained in the transverse direction of the geosynthetic reinforcement when summing the two orthogonal layers together, see also Table 2.6 and Chapter 8.3.1.

The same applies to adding up the stiffnesses in the two directions. If the transverse and longitudinal reinforcements nonetheless have different properties, account should be taken of differences in the force-strain behaviour; see Fig. 4.4.

Table 4.1 Determination of the GR tensile stiffness based on tensile force/strain isochronous curves (principle).

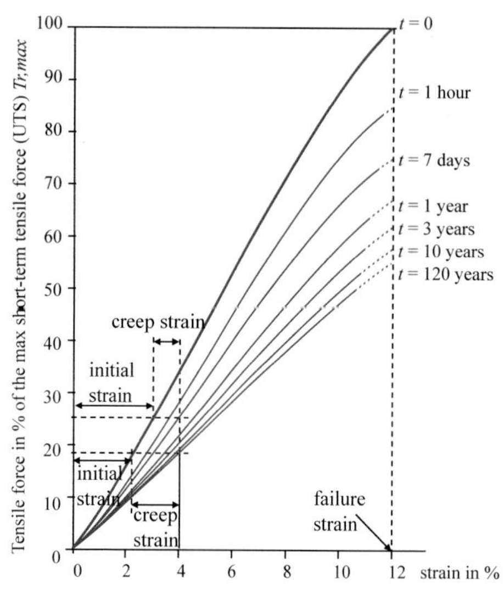

Max. short-term tensile strength of reinforcement $T_{r;max}$ = 200 kN/m	
Estimated SLS strain, ε = 4%	
Construction phase (t = 7 days)	Service life (t = 120 years)
Read off % of loading = 25%	Read off % of loading = 18%
Tensile stiffness, $J_{t=7days}$ $J = T_{r;max} \cdot 25\% / \varepsilon$ $= 200 \cdot 25\% / 4\% =$ 1250 kN/m	Tensile stiffness, $J_{t=120yrs}$ $J = T_{r;max} \cdot 18\% / \varepsilon$ $= 200 \cdot 18\% / 4\% =$ 900 kN/m
N.B.: The isochronous curves shown here are examples only. For design, the curves that belong with the specific geosynthetic reinforcement to be employed should be used. The geosynthetic reinforcement manufacturer can provide these. Chapter 5.5 gives a calculation example.	

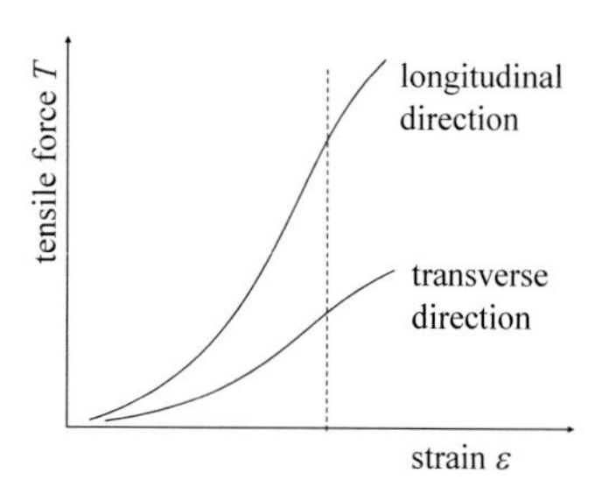

Fig. 4.4 Superposition of stiffness in longitudinal and transverse direction of geosynthetic reinforcement.

4.3.2.4 Traffic load

The traffic load should be determined in accordance with Chapter 2.3.

4.3.2.5 Boundary conditions for application of design method for geosynthetic reinforcement

In the design and dimensioning of the reinforced embankment the boundary conditions prescribed in Table 4.2 should be met. When deviations occur, it should be demonstrated by means of additional calculations and/or suitability tests that the structure meets the requirements imposed.

Table 4.2 Boundary conditions for application of the design method for the geosynthetic reinforcement.

No.	Boundary condition
1	For the ratio between the embankment height H and the diagonal net spacing between the pile caps, the following should apply: $H / (s_d - d_{eq}) \geq 0.66$
2	The weight of the embankment should be such that the dynamic load p is less than or equal to 50% of the total load $\sigma_{v;tot}$ (static + dynamic): $p / \sigma_{v;tot} \leq 0.50$. If this ratio becomes greater, the reduction factor κ of Heitz (2006) should be applied; see Chapter 2.5.3.
3	For the ratio between pile cap width b and centre-to-centre pile spacing $s_{x,y}$, the following should apply: $b / s_{x,y} \geq 0.15$
4	For a single reinforcement layer (single biaxial or two uniaxial orthogonal to each other), or the bottom layer in case of more than one GR layer, the distance between pile cap and reinforcement layer z should be $z \leq 0.15$ m. With two reinforcement layers, the distance between the two layers of reinforcement should be no more than 0.20 m.
5	The ratio between the centre-to-centre spacing of the piles in the orthogonal directions should be: $2/3 \leq s_x / s_y \leq 3/2$.
6	Within the range of the mattress height $h^* = 0.66\ (s_d - d_{eq})$ above the pile cap, the fill should be composed of granular material with a characteristic internal friction angle φ'_{cv} of at least 35° (this is the residual value). In the embankment above h^*, it applies that $\varphi'_{cv} \geq 30°$.
7	The minimum GR tensile strength to be used in the structure should be $T_{r;d} = 30$ kN/m in both orthogonal directions for each reinforcement layer. The ratio between the tensile strength of each reinforcement layer in both directions should meet: $0.1 \leq T_{r;x;d} / T_{r;y;d} \leq 10$
8	The ratio between the values of subgrade reaction for the piles and the subsoil should meet: $k_{s;pile} / k_{s;subsoil} > 10$

During construction of the embankment the boundary conditions stated above may not all be met. This obviously applies to the requirement stated regarding embankment height. During the construction phase, only a small number of traffic passages will occur, and further, support from the subsoil can usually be relied upon. Thus, the design method can be considered also applicable for the calculations in the construction phase.

4.3.2.6 Design in two steps

The following are derived using the design method described in this Chapter:

- the consequences of the vertical load (Chapter 4.3.2);
  - calculation step 1, arching: divides the vertical load into two parts (see also Fig. 2.1);
    - total force on the geosynthetic reinforcement between the piles, *B+C* in kN/pile (or kN/pile unit);
    - total vertical force on the pile caps, *A* in kN/pile;
  - calculation step 2, calculation of strain and tensile force;
    - the maximum GR tensile force and GR strain as a result of the vertical force *B+C* from calculation step 1;
- the consequences of the horizontal load are described in Chapters 2.4, 4.3.3 and 4.3.4.

The calculation described below is carried out using design values of the various material parameters (soil and reinforcement) and loads. Chapter 5 gives calculation examples for the design of the geosynthetic reinforcement.

An excel file containing the basic equations of the Concentric Arches model is available for download at **www.piledembankments.com** or at **www.crcpress.com/9789053676240**.

Calculation step 1: division of vertical load; calculation of vertical load on the geosynthetic reinforcement between the piles (*B+C*)

The average vertical stress $\sigma_{v;tot}$ (kPa) at the reinforcement level, neglecting the arching, is:

$$\sigma_{v;tot} = \Sigma \gamma \cdot H + p \tag{4.6}$$

where:

$\sigma_{v;tot}$	kPa	the average vertical stress at the level of the geosynthetic reinforcement, neglecting arching,
p	kPa	the traffic load on the road surface, distributed to the level of the geosynthetic reinforcement (obtained from Table 2.2, Table 2.3 or a table in Appendix A). Any permanent load is added to this, so $p = p_{traffic} + p_{permanent}$. Note the instructions in Chapter 2.4 concerning braking and centrifugal forces. This is detailed in the calculation examples (Cases 1a and 1b) in Chapter 5,
γ	kN/m³	unit weight of the fill,
H	m	embankment height (distance between top of pile cap and top of asphalt or track bed).

The total force F_{tot} in kN/pile, so per pile unit ($s_x \cdot s_y$) is:

$$F_{tot} = A + B + C = (\Sigma\, \gamma \cdot H + p) \cdot s_x \cdot s_y \tag{4.7}$$

Due to arching, the vertical pressure on the geosynthetic reinforcement between the piles will be lower than that above the pile caps. Thus, A will be greater than ($\sigma_{v;tot} \cdot A_p$), and B+C will be less than ($\sigma_{v;tot} \cdot (A_i - A_p)$), where:

A_i	m^2	the area of influence of the pile = $s_x \cdot s_y$,
A_p	m^2	the area of the top of the pile cap.

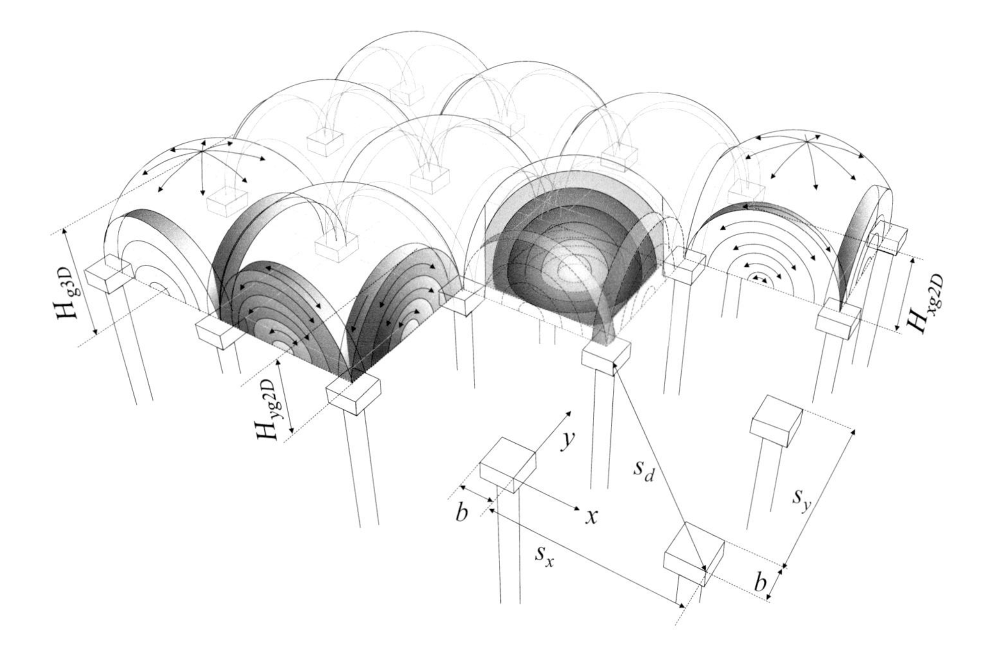

Fig. 4.5 Concentric Arches model of Van Eekelen et al., 2013 [21]

The magnitude of A and B+C (kN/pile) are calculated as follows using the Concentric Arches model of Van Eekelen *et al.* (2013) [21], see Fig. 4.5.

$$\begin{aligned} &\left(B+C\right)^{stat}_{p=0} = F_{GRsquare;p=0} + F_{GRstrip;p=0} \quad \text{and} \\ &A^{stat}_{p=0} = \left(\gamma H\right) \cdot s_x \cdot s_y - F_{GRsquare;p=0} - F_{GRstrip;p=0} \end{aligned} \tag{4.8}$$

Emerging from the equations given later, the following is generated:

$$(B+C)_{p>0}^{stat} = \left(\frac{\gamma H + p}{\gamma H}\right) \cdot (B+C)_{p=0}^{stat}$$

$$A_{p>0}^{stat} = \left(\frac{\gamma H + p}{\gamma H}\right) \cdot A_{p=0}^{stat} \tag{4.9}$$

If $\frac{p_{traffic}}{\sigma_{v;tot}} \leq 0.50$ (see Chapter 2.5.3) then equation (4.8) and (4.9) are applicable (κ = 1.0 in equation (4.10)), where $p_{traffic}$ is the traffic load on the surface, obtained from Table 2.2, Table 2.3 or the tables in Appendix A (in kPa) and in which a permanent load that may be present is not included.

Alternatively, if $\frac{p_{traffic}}{\sigma_{v;tot}} > 0.50$, $(B+C)^{cycl}$ and A^{cycl} in equation (4.10) should be used:

$$(B+C)_{p>0}^{cycl} = (\gamma \cdot H + p) \cdot s_x \cdot s_y \cdot \left(1 - \frac{1}{\kappa}\right) + \frac{(B+C)_{p>0}^{stat}}{\kappa}$$

$$A_{p>0}^{cycl} = (\gamma H + p) \cdot s_x \cdot s_y - (B+C)_{p>0}^{cycl} \tag{4.10}$$

$(B+C)^{stat}$	kN/pile	the total vertical force at construction level on the geosynthetic reinforcement between the pile caps in kN/pile; see Fig.2.1, omitting the effect of dynamic load,
$(B+C)^{cycl}$	kN/pile	the total vertical force at construction level on the geosynthetic reinforcement between the pile caps in kN/pile; see Fig.2.1, where the value of $(B+C)^{cycl}$ is greater than that of $(B+C)^{stat}$, so that the effect of a relatively large dynamic load is taken into account. By this means the arching is reduced,
κ	-	the factor for dynamic load according to Heitz [16]; see Fig. 4.6.

$F_{GRsquare}$ (kN/pile) is determined as follows:

$$F_{GRsquare;p>0} = \left(\frac{\gamma H + p}{\gamma H}\right) \cdot F_{GRsquare;p=0}$$

where

$$F_{GRsquare;p=0} = F_{GRsq1\,p=0} + F_{GRsq2\,p=0} + F_{GRsq3\,p=0}$$

where

$$F_{GRsq1\,p=0} = \frac{\pi P_{3D}}{K_p} \cdot \left(\frac{L_{3D}}{2}\right)^{2K_p} + \tfrac{2}{3}\pi Q_{3D} \cdot \left(\frac{L_{3D}}{2}\right)^3$$

$$F_{GRsq2\,p=0} = {}_1F_{GRsq2} + {}_2F_{GRsq2} + {}_3F_{GRsq2} + {}_4F_{GRsq2}$$

where

$${}_1F_{GRsq2} = \frac{2\pi P_{3D}}{2K_p}\left(2^{K_p} - 1\right)\left(\frac{L_{3D}}{2}\right)^{2K_p} \tag{4.11}$$

$${}_2F_{GRsq2} = \frac{2\pi Q_{3D}}{3}\left(\sqrt{2}^3 - 1\right)\left(\frac{L_{3D}}{2}\right)^3$$

$${}_3F_{GRsq2} = \frac{P_{3D} \cdot 2^{2-2K_p} \cdot L_{3D}^{2K_p}}{K_p} \cdot \left(-\frac{\pi}{2^{2-K_p}} + \sum_{n=0}^{\infty} \frac{1}{2n+1}\binom{K_p - 1}{n}\right)$$

$$= \frac{P_{3D} \cdot 2^{2-2K_p} \cdot L_{3D}^{2K_p}}{K_p} \cdot \left(\begin{array}{l} -\frac{\pi}{2^{2-K_p}} + 1 + \frac{1}{3}\left(K_p - 1\right) + \frac{1}{10}\left(K_p - 1\right)\left(K_p - 2\right) \\ + \frac{1}{42}\left(K_p - 1\right)\left(K_p - 2\right)\left(K_p - 3\right) \\ + \frac{1}{216}\left(K_p - 1\right)\left(K_p - 2\right)\left(K_p - 3\right)\left(K_p - 4\right) + \frac{1}{1320}....\left(K_p - 5\right)... \end{array}\right)$$

$${}_4F_{GRsq2} = \tfrac{1}{6} Q_{3D} L_{3D}^{\ 3} \cdot \left(\sqrt{2}\left(1 - \pi\right) + \ln\left(1 + \sqrt{2}\right)\right)$$

where:

$$P_{3D} = \gamma \cdot K_p \cdot H_{g3D}{}^{2-2K_p} \cdot \left[H - H_{g3D} \cdot \left(\frac{2K_p - 2}{2K_p - 3} \right) \right] \text{ and} \tag{4.12}$$

$$Q_{3D} = K_p \cdot \frac{\gamma}{2K_p - 3}$$

And where H_{g3D} (in m) is the height of the largest hemisphere (see Fig. 4.5):

$$H_{g3D} = \frac{s_d}{2} \quad \text{for} \quad H \geq \frac{s_d}{2} \qquad \text{(full arching)}$$

(4.13)

$$H_{g3D} = H \quad \text{for} \quad H < \frac{s_d}{2} \qquad \text{(partial arching)}$$

L_{3D} is given by:

$$L_{3D} = \frac{1}{\sqrt{2}} \sqrt{(s_x - b_{eq})^2 + (s_y - b_{eq})^2} \quad \text{for } H \geq \frac{1}{2} \sqrt{(s_x - b_{eq})^2 + (s_y - b_{eq})^2}$$

(4.14)

$$L_{3D} = \sqrt{2} \cdot H_{g3D} \quad \text{for } H < \frac{1}{2} \sqrt{(s_x - b_{eq})^2 + (s_y - b_{eq})^2}$$

The passive earth pressure coefficient K_p is given by:

$$K_p = \frac{1 + \sin \varphi}{1 - \sin \varphi} \tag{4.15}$$

For shallow embankments, $F_{GRsq3\,p=0}$ results in a value greater than zero:

$$F_{GRsq3\,p=0} = \gamma H \cdot \left((s_x - b_{eq}) \cdot (s_y - b_{eq}) - L_{3D}^2 \right) \text{ for } L_{3D}^2 < (s_x - b_{eq}) \cdot (s_y - b_{eq})$$

(4.16)

$$F_{GRsq3\,p=0} = 0 \quad \text{for } L_{3D}^2 \geq (s_x - b_{eq}) \cdot (s_y - b_{eq})$$

The load that is transferred from the 3D hemispheres to the 2D arches is equal to:

$$F_{transferred} = \gamma H \cdot (s_x - b_{eq}) \cdot (s_y - b_{eq}) - (F_{GRsq1\,p=0} + F_{GRsq2\,p=0} + F_{GRsq3\,p=0})$$
$$p_{transferred} = \frac{F_{transferred}}{b_{eq} \cdot (L_{x2D} + L_{y2D}) + b_{eq}^2} \tag{4.17}$$

where L_{x2D} and L_{y2D} are given by:

$$\begin{aligned}
L_{x2D} &= s_x - b_{eq} && \text{for } H \geq \tfrac{1}{2}(s_x - b_{eq}) \\
L_{x2D} &= 2 \cdot H_{xg2D} && \text{for } H < \tfrac{1}{2}(s_x - b_{eq}) \\
L_{y2D} &= s_y - b_{eq} && \text{for } H \geq \tfrac{1}{2}(s_y - b_{eq}) \\
L_{y2D} &= 2 \cdot H_{yg2D} && \text{for } H < \tfrac{1}{2}(s_y - b_{eq})
\end{aligned} \tag{4.18}$$

And where the height of the 2D arches (Fig. 4.5) is given by:

$$\begin{aligned}
H_{xg2D} &= \frac{s_x}{2} && \text{for } H \geq \frac{s_x}{2} \\
H_{xg2D} &= H && \text{for } H < \frac{s_x}{2} \\
H_{yg2D} &= \frac{s_y}{2} && \text{for } H \geq \frac{s_y}{2} \\
H_{yg2D} &= H && \text{for } H < \frac{s_y}{2}
\end{aligned} \tag{4.19}$$

$F_{GRstrip}$ (kN/pile) is determined as follows:

$$F_{GRstrip;p>0} = \left(\frac{\gamma H + p}{\gamma H}\right) \cdot F_{GRstrip;p=0} \tag{4.20}$$

where

$$F_{GRstrip;p=0} = 2b_{eq}\frac{P_{x2D}}{K_p}\left(\tfrac{1}{2}L_{x2D}\right)^{K_p} + \tfrac{1}{4}b_{eq}Q_{2D} \cdot \left(L_{x2D}\right)^2 + F_{xGRstr2\,p=0}$$

$$+2b_{eq}\frac{P_{y2D}}{K_p}\left(\tfrac{1}{2}L_{y2D}\right)^{K_p} + \tfrac{1}{4}b_{eq}Q_{2D} \cdot \left(L_{y2D}\right)^2 + F_{yGRstr2\,p=0}$$

where:

$$P_{x2D} = K_p \cdot H_{xg2D}^{\left(1-K_p\right)} \cdot \left[\gamma H + p_{transferred} - \gamma H_{xg2D} \cdot \left(\frac{K_p - 1}{K_p - 2}\right)\right]$$

$$P_{y2D} = K_p \cdot H_{yg2D}^{\left(1-K_p\right)} \cdot \left[\gamma H + p_{transferred} - \gamma H_{yg2D} \cdot \left(\frac{K_p - 1}{K_p - 2}\right)\right]$$

$$Q_{2D} = K_p \cdot \frac{\gamma}{K_p - 2} \tag{4.21}$$

$$F_{xGRstr2\,p=0} = \gamma H b_{eq}\left(s_x - b_{eq} - L_{x2D}\right) \qquad \text{for } H < \tfrac{1}{2}\left(s_x - b_{eq}\right)$$

$$F_{xGRstr2\,p=0} = 0 \qquad \text{for } H \geq \tfrac{1}{2}\left(s_x - b_{eq}\right)$$

$$F_{yGRstr2\,p=0} = \gamma H b_{eq}\left(s_y - b_{eq} - L_{y2D}\right) \qquad \text{for } H < \tfrac{1}{2}\left(s_y - b_{eq}\right)$$

$$F_{yGRstr2\,p=0} = 0 \qquad \text{for } H \geq \tfrac{1}{2}\left(s_y - b_{eq}\right)$$

When $\kappa \geq 1.0$, $(B{+}C)^{cycl}$ becomes greater than $(B{+}C)^{stat}$. Accordingly, load part A which transfers directly to the piles becomes less and the load on the remaining surface area (geosynthetic B + subsoil C) becomes greater for a relatively larger dynamic load. Consequently, the arching is reduced. The factor κ may be derived from Fig. 4.6. Typical cyclic frequencies for trains and road traffic are listed in Table 4.3.

Table 4.3 Cyclic frequencies for railway and road traffic.

Speed		Railway lines *) Train axle spacing	Railway lines *) Train impact frequency	Roads **) Truck impact frequency
km/h	m/s	m	Hz	Hz
40	11	11	1	0.5 – 1.0
80	22	11	2	0.5 – 1.0
120	33	11	3	0.5 – 1.0
160	44	11	4	0.5 – 1.0

**) This is for a cargo train*

***) For road traffic, the frequency is calculated based on the safety requirement to maintain two seconds' separation from the vehicle in front.*

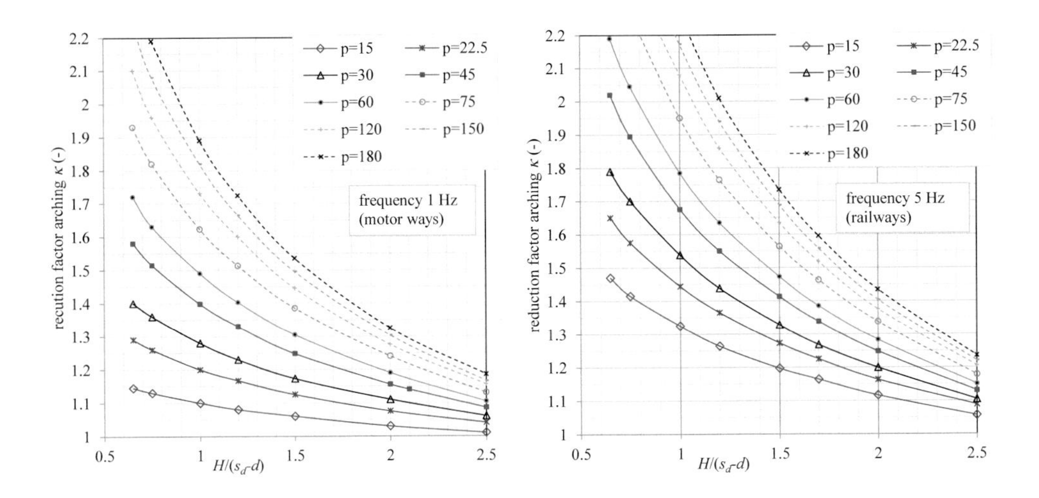

Fig. 4.6 Reduction factor κ according to Heitz (after Fig. 8.3b on p. 157 of Heitz (2006) [16]), where $p = p_{traffic}$ in kPa, obtained from Table 2.2, Table 2.3 or Tables A.1 to A.8 in Appendix A. The broken lines above have been added to Heitz's original curves. They are the results of extrapolation by the authors. No increase in the traffic load as a result of braking forces or centrifugal forces is included in the determination of the arching reduction factor κ.

Total load on the pile caps (A) and (T_{max})

The vertical force (kN/pile) that is imposed on the pile caps via arching is given by equations (4.8) and (4.10). The average stress acting on the piles via arching is thus:

$$\sigma_{v;p}^{stat} = \frac{A^{stat}}{A_p} = \frac{(\gamma H + p) \cdot A_i - F_{GRsquare} - F_{GRstrip}}{A_p} \tag{4.22}$$

$$\sigma_{v;p}^{cycl} = \frac{A^{cycl}}{A_p} = \frac{(\gamma H + p) \cdot A_i - (B + C)^{cycl}}{A_p} \tag{4.23}$$

where:

$\sigma_{v;p}$	kPa	the vertical stress as a result of arching acting on the pile caps (above the GR).

If $\frac{p}{\sigma_{v;tot}} \leq 0.50$ (see Chapter 2.5.3), $\sigma_{v;p}^{stat}$ applies (equation (4.22)), otherwise $\sigma_{v;p}^{cycl}$ applies (equation (4.23)).

The pile load calculated using equations (4.22) and (4.23) is the load that is transferred to the piles by arching alone. This may be assumed as an evenly distributed load on the pile cap for the purposes of design of the pile cap. To this needs to be added the GR tensile force $T_{max,x}$ (kN/m) and $T_{max,y}$ (kN/m) (see Chapter 3.3). The magnitude of this is dependent on the amount of subsoil support between the piles.

For the dimensioning of the piles (structural check), it is assumed that the subsoil does not contribute, with the pile load (kN) calculated as follows:

$$F_p = (\Sigma\, \gamma \cdot H + p) \cdot A_i = \sigma_{v;tot} \cdot A_i \tag{4.24}$$

Calculation step 2: calculation of the GR tensile forces and GR strains
The stresses on the geosynthetic reinforcement, calculated in the previous Chapter, are converted to loads acting on the reinforcement strips between each pair of adjacent pile caps. Here it is assumed that the reinforcement strips on which the loads act are oriented in orthogonal directions (x and y) in the directions in which the reinforcement layers can absorb tensile forces, and have a width that is equal to the equivalent width b_{eq} of the pile cap.

An area of influence A_{Lx} or A_{Ly} is defined for each strip as indicated in Fig. 4.7. These areas of influence may be calculated using the equations below. The part that is situated above the pile caps is substracted from the total area of the diamond that contains the area of influence.

$$A_{Lx} = \frac{1}{2} \cdot \left(s_x \cdot s_y\right) - \frac{d^2}{2} \cdot \arctan\left[\frac{s_y}{s_x}\right] \quad (4.25)$$

$$A_{Ly} = \frac{1}{2} \cdot \left(s_x \cdot s_y\right) - \frac{d^2}{2} \cdot \arctan\left[\frac{s_x}{s_y}\right] \quad (4.26)$$

where the arctan should be calculated in radians.

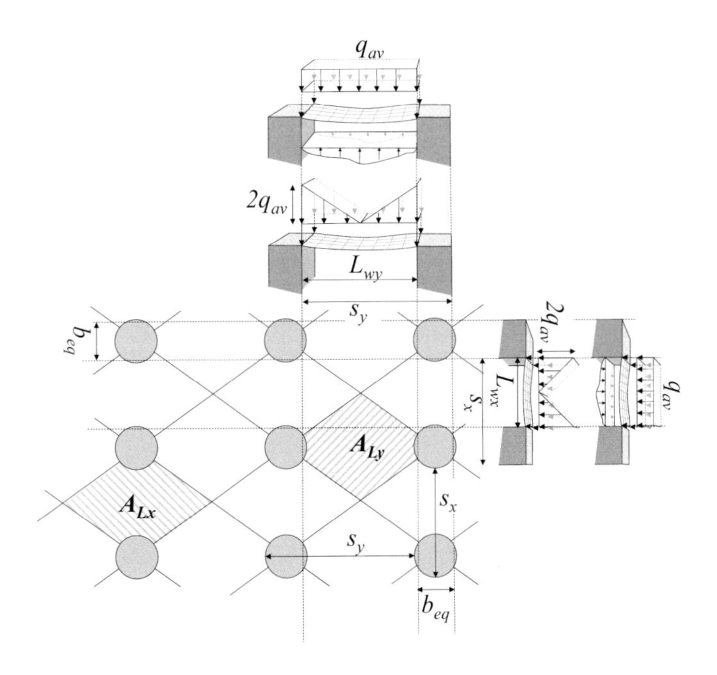

Fig. 4.7 Areas of influence of the GR strips between the piles; these are used for determining the modified subgrade reaction according to equations (4.29) and (4.30).

The length of the GR strips is the net spacing L_w between the piles:

$$L_{wx} = s_x - b_{eq} \qquad L_{wy} = s_y - b_{eq} \quad (4.27)$$

The average vertical stress q_{av} (kPa) that acts on the GR strips is calculated by dividing the total force on the reinforcement by the sum of the areas of the reinforcement strips in the transverse and longitudinal directions:

$$q_{av} = \frac{B+C}{b_{eq}\left(L_{wx}+L_{wy}\right)} \tag{4.28}$$

The subsoil support beneath the reinforcement strips is taken into account using a modified subgrade reaction value K (kN/m^3) in which the support of the entire areas A_{Lx} and A_{Ly} are taken into account:

$$K_x = \frac{A_{Lx} \cdot k_s}{L_{wx} \cdot b_{eq}} \tag{4.29}$$

$$K_y = \frac{A_{Ly} \cdot k_s}{L_{wy} \cdot b_{eq}} \tag{4.30}$$

The resulting GR strains may be found using the equations below, or using the design charts in Fig. 4.9 to Fig. 4.12. The GR strains are determined for the GR strips between two adjacent pile caps. The load distribution on the GR strips may be uniform (*uni*) or inverse-triangular (*inv*) as indicated in Fig. 4.7. The maximum GR strain is determined for both load distributions ($\varepsilon_{inv,max}$ and $\varepsilon_{uni,max}$). This maximum strain occurs near the edge of the pile caps. Then the minimum value of these two strains is used as the controlling strain level. So, the controlling strain is min($\varepsilon_{inv,max}$; $\varepsilon_{uni,max}$). This approach is described in Van Eekelen *et al.* (2012b and 2015, [20] and [22]) and shown in the calculation examples of Chapter 5.5.

Under vertical load the GR strips will deflect. This deflection is necessary to allow the arching to develop. The shape of the deflected reinforcement strip $z(x)$, and the derivative of this $z'(x)$ are described by the equations in the following table. These may be programmed into a spreadsheet program for example, where the reinforcement strips are divided up into small increments Δx.

From equation (4.31) to equation (4.38), all values should be calculated for both the uniform and the inverse-triangular load distribution.

Load distribution	**GR deflection *z*(*x*) and its derivative *z*'(*x*)**	
Uniform	$z_{uni}(x) = -\frac{q_{av}}{K}\left(\frac{e^{\alpha x}+e^{-\alpha x}}{e^{\frac{1}{2}\alpha L_w}+e^{-\frac{1}{2}\alpha L_w}}-1\right)$ $z'_{uni}(x) = -\frac{q_{av}\alpha}{K}\left(\frac{e^{\alpha x}-e^{-\alpha x}}{e^{\frac{1}{2}\alpha L_w}+e^{-\frac{1}{2}\alpha L_w}}\right)$ Without subsoil support (K=0 kN/m^3): $z_{uni}(x) = -\frac{q_{av}L_w^2}{8T_H}\left(4\left(\frac{x}{L_w}\right)^2-1\right)$ $z'_{uni}(x) = \frac{q_{av}}{T_H}x$	(4.31)
Inverse triangle 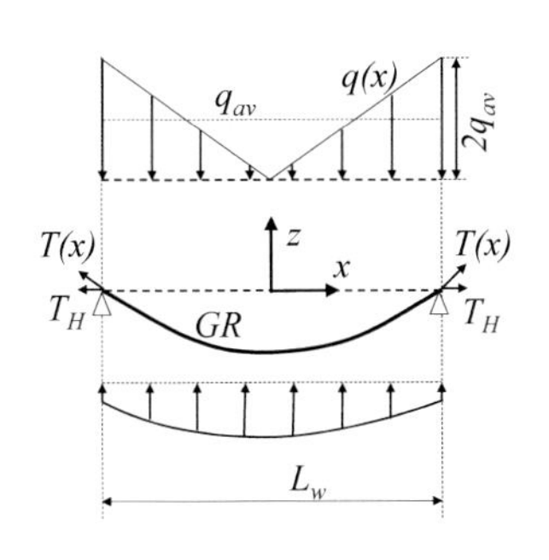 	$z_{inv}(x) = -\frac{2q_{av}}{KL_w\alpha}\left(Me^{\alpha x}+\left(M-2\right)e^{-\alpha x}-2\alpha x\right)$ $z'_{inv}(x) = -\frac{2q_{av}}{KL_w}\left(Me^{\alpha x}-\left(M-2\right)e^{-\alpha x}-2\right)$ where $M = \frac{L_w\alpha+2e^{-\frac{1}{2}\alpha L_w}}{e^{\frac{1}{2}\alpha L_w}+e^{-\frac{1}{2}\alpha L_w}}$ Without subsoil support (K=0 kN/m^3): $z_{inv}(x) = -\frac{q_{av}L_w^2}{12T_H}\left(8\left(\frac{x}{L_w}\right)^3-1\right)$ $z'_{inv}(x) = -\frac{2q_{av}L_w}{T_H}\cdot\left(\frac{x}{L_w}\right)^2$	(4.32)

In the above equations, α is given by:

$$\alpha^2 = \frac{K}{T_H} \tag{4.33}$$

T_H (kN/m) is the horizontal component of the GR tensile force. The value of T_H is initially unknown and is solved for below.

The GR tensile force $T(x)$ in kN/m is calculated as follows:

$$T(x) = T_H \sqrt{1 + \left(z'(x)\right)^2} \tag{4.34}$$

The average geometric strain is calculated using:

$$\varepsilon_{geometric,average} = \frac{\int_{x=0}^{x=\frac{1}{2}L} dx \sqrt{1 + \left(z'(x)\right)^2} - \frac{1}{2}L}{\frac{1}{2}L} \tag{4.35}$$

Where $z'(x)$ is given by equations (4.31) or (4.32) and the average constitutive strain by:

$$\varepsilon_{const,average} = \frac{\frac{1}{J} \int_{x=0}^{x=\frac{1}{2}L} T(x)\, dx}{\frac{1}{2}L} \tag{4.36}$$

Where $T(x)$ is given by equation (4.34). The value of the unknown T_H is found by equalising the average geometrical and the average constitutive strain:

$$\varepsilon_{geom,average} = \varepsilon_{const,average} \tag{4.37}$$

By this process the distribution of the GR tensile force $T(x)$ and the GR strain $\varepsilon(x)$ in the reinforcement strip is found. The maximum GR strain and GR tensile force are found at the edge of the pile cap, so for $x = L_w/2$:

$$T_{max} = \mathrm{T}\left(\frac{L_w}{2}\right) = T_H \sqrt{1 + \left(z'\left(\frac{L_w}{2}\right)\right)^2} \tag{4.38}$$

$$\varepsilon_{max} = \frac{T_{max}}{J}$$

The calculations are conducted to this point for both the uniform and the inverse-triangular load distribution cases. The results obtained are the tensile forces $T_{inv,\max}$ and $T_{uni,\max}$ and the GR strains $\varepsilon_{inv,\max}$ and $\varepsilon_{uni,\max}$ and the associated GR deflection. The minimum of the two sets of results gives the appropriate design values:

$$\varepsilon_{g,\max,x} = \min(\varepsilon_{inv,\max,x} \,;\, \varepsilon_{uni,\max,x}) \text{ and } \varepsilon_{g,\max,y} = \min(\varepsilon_{inv,\max,y} \,;\, \varepsilon_{uni,\max,y})$$

$$T_{\max,x} = \min(T_{inv,\max,x} \,;\, T_{uni,\max,x}) \text{ and } T_{\max,y} = \min(T_{inv,\max,y} \,;\, T_{uni,\max,y}) \tag{4.39}$$

For the normative load distribution (uniform or inverse triangle), the associated maximum GR deflection is determined. This maximum often (but not always) lies in the centre between the pile caps.

$$f_{\max,\mathrm{GRstrip}\_x,y} = \max\left(z(x)\right) \quad \text{for} \quad x = 0 \quad \text{to} \quad x = \frac{L_w}{2} \tag{4.40}$$

Fig. 4.8 gives a calculation procedure that may be followed to program a spreadsheet to calculate the GR strains.

Determination maximal strain in GR strip // x -direction

(1) Geometry, properties and load

		// x-direction	⊥ y-direction
centre-to-centre distance	m	s_x	s_y
equivalent width of pile caps	m	$b_{eq} = \sqrt{A_p}$	$b_{eq} = \sqrt{A_p}$
GR stiffness	kN/m	J_x	J_y

(2) Calculated results

net spacing between pile caps	m	L_{wx}	L_{wy}
area GR strip	m^2	$A_{GRstrip,x} = L_{wx} \cdot b_{eq}$	$A_{GRstrip,y} = L_{wy} \cdot b_{eq}$
area belonging to GR strip (Fig 4.7)	m^2	A_{Lx}	A_{Ly}
average load q_{av} on GR strips (eq. 4.28)	kPa	$q_{av} = (B+C)^{cycl} / (A_{GRstrip,x} + A_{GRstrip,y})$	$q_{av} = (B+C)^{cycl} / (A_{GRstrip,x} + A_{GRstrip,y})$

(3a) Invers-triangular load distribution (example with 50 increments)

x	$z_{inv}(x)$	$z'_{inv}(x)$	$T_{inv}(x)$	$\varepsilon_{inv}(x)$	$z'_{inv,average}(x)$	$T_{inv,average}(x)$
Equation number	4.32	4.32 in end point increment	4.34 in end point increment	in end point increment	average in increment (xi, xi+1)	average in increment (xi, xi+1)
$x0=0$	$z_{inv}(x0)$	$z'_{inv}(x0)$	$T_{inv}(x0)$	$T_{inv}(x0)/J$	$average(z'_{inv}(x0); z'_{inv}(x1))$	$average(T_{inv}(x0); T_{inv}(x1))$
$x1=L_{Wx}/100$	$z_{inv}(x1)$	$z'_{inv}(x1)$	$T_{inv}(x1)$	$T_{inv}(x1)/J$	$average(z'_{inv}(x1); z'_{inv}(x2))$	$average(T_{inv}(x1); T_{inv}(x2))$
$x2=x1+L_{Wx}/100$	$z_{inv}(x2)$	$z'_{inv}(x2)$	$T_{inv}(x2)$	$T_{inv}(x2)/J$	$average(z'_{inv}(x2); z'_{inv}(x3))$	$average(T_{inv}(x2); T_{inv}(x3))$
$x3=x2+L_{Wx}/100$	$z_{inv}(x3)$	$z'_{inv}(x3)$	$T_{inv}(x3)$	$T_{inv}(x3)/J$	$average(z'_{inv}(x3); z'_{inv}(x4))$	$average(T_{inv}(x3); T_{inv}(x4))$
.....	....	....	....	....	....	....
$x49=L_{wx}/2-L_{Wx}/100$	$z_{inv}(x49)$	$z'_{inv}(x49)$	$T_{inv}(x49)$	$T_{inv}(x49)/J$	$average(z'_{inv}(x49); z'_{inv}(x50))$	$average(T_{inv}(x49); T_{inv}(x50))$
$x50=L_{wx}/2$	$z_{inv}(x50)$	$z'_{inv}(x50)$	$T_{inv}(x50)$	$T_{inv}(x50)/J$		

(3b) Uniform load distribution (example with 50 increments)

x	$z_{uni}(x)$	$z'_{uni}(x)$	$T_{uni}(x)$	$\varepsilon_{uni}(x)$	$z'_{uni,average}(x)$	$T_{uni,average}(x)$
equation number	4.31	4.31 in end point increment	4.34 in end point increment	in end point increment	average in increment (xi, xi+1)	average in increment (x i, xi+1)
$x0=0$	$z_{uni}(x0)$	$z'_{uni}(x0)$	$T_{uni}(x0)$	$T_{uni}(x0)/J$	$average(z'_{uni}(x0); z'_{uni}(x1))$	$average(T_{uni}(x0); T_{uni}(x1))$
$x1=L_{Wx}/100$	$z_{uni}(x1)$	$z'_{uni}(x1)$	$T_{uni}(x1)$	$T_{uni}(x1)/J$	$average(z'_{uni}(x0); z'_{uni}(x1))$	$average(T_{uni}(x0); T_{uni}(x1))$
$x2=x1+L_{Wx}/100$	$z_{uni}(x2)$	$z'_{uni}(x2)$	$T_{uni}(x2)$	$T_{uni}(x2)/J$	$average(z'_{uni}(x0); z'_{uni}(x1))$	$average(T_{uni}(x0); T_{uni}(x1))$
$x3=x2+L_{Wx}/100$	$z_{uni}(x3)$	$z'_{uni}(x3)$	$T_{uni}(x3)$	$T_{uni}(x3)/J$	$average(z'_{uni}(x0); z'_{uni}(x1))$	$average(T_{uni}(x0); T_{uni}(x1))$
.....	....	....	....	....	....	....
$x49=L_{wx}/2-L_{Wx}/100$	$z_{uni}(x49)$	$z'_{uni}(x49)$	$T_{uni}(x49)$	$T_{uni}(x49)/J$	$average(z'_{uni}(x0); z'_{uni}(x1))$	$average(T_{uni}(x0); T_{uni}(x1))$
$x50=L_{wx}/2$	$z_{uni}(x50)$	$z'_{uni}(x50)$	$T_{uni}(x50)$	$T_{uni}(x50)/J$		

(4) Determine a value T_H that give equal geometric and constitutive strain

	Inverse triangular load distribution:	**uniform load distribution:**
	change value of T_{H_inv} until $\varepsilon_{inv,geom} - \varepsilon_{inv,const} = 0$	change value of T_{H_uni} until $\varepsilon_{uni,geom} - \varepsilon_{uni,const} = 0$
horizontal component tensile force T	T_{H_inv}	T_{H_uni}
average geometric GR strain (eq. 4.35)	$\varepsilon_{inv,geom,average}$	$\varepsilon_{uni,geom,average}$
average constitutive GR strain (eq. 4.36)	$\varepsilon_{inv,const,average}$	$\varepsilon_{uni,const,average}$
difference geometric and constitutive GR strain	$\varepsilon_{inv,geom,average} - \varepsilon_{inv,const,average}$	$\varepsilon_{uni,geom,average} - \varepsilon_{uni,const,average}$

(5) Results: GR strain, tensile force, GR deflection

resulting average GR strain	$min(\varepsilon_{inv,average}; \varepsilon_{uni,average})$
thus normative is	inverse triangular or uniformly distributed load
resulting max GR strain	$max(\varepsilon_{normative\ load\ distribution\ in\ increment\ end\ points})$
resulting max GR tensile force	$max(T_{normative\ load\ distribution\ in\ increment\ end\ points})$
resulting max GR deflection	$max(z_{normative\ load\ distribution})$

Fig. 4.8 Example design procedure for calculation step 2; see also the design charts in Fig. 4.9 to Fig. 4.12

Using the design charts in Fig. 4.9 to Fig. 4.12, the maximum GR strain ε_{max} can also be read as a function of the following dimensionless parameters:

$$\frac{q_{av} \cdot L_w}{J} \qquad \text{plot on the } x\text{-axis} \tag{4.41}$$

$$\frac{k_S \cdot A_{Lx,y} \cdot L_w}{J \cdot b_{eq}} \qquad \text{plot in curves} \tag{4.42}$$

The maximum GR strain $\varepsilon_{uni,max}$ for the uniform load distribution may be found in Fig. 4.9 or Fig. 4.10. Here, Fig. 4.10 is a detailed part of Fig. 4.9. The maximum GR strain $\varepsilon_{inv,max}$ for the inverse-triangular load distribution may be found in Fig. 4.11 or Fig. 4.12. Here, Fig. 4.12 is a detailed part of Fig. 4.11. The load distribution that gives the lesser GR strain ε_{max} should be used for design.

The maximum GR tensile force may then be calculated from the design strain:

$$T_{v;max} = \varepsilon_{max} \cdot J \tag{4.43}$$

where:

$T_{v;max}$	kN/m	the maximum GR tensile force as a result of the vertical load,
ε_{max}	-	the maximum GR strain,
J	kN/m	the GR tensile stiffness, see Chapter 4.3.2.3.

The equations above apply to both the x and y directions. A GR tensile force is found in both the x and y directions. The parameters L_w and J and thus ε and T depend on the direction (x or y).

Van Eekelen *et al.* (2012b and 2015) provided the derivation of the design charts in Fig. 4.9 to Fig. 4.12 in detail.

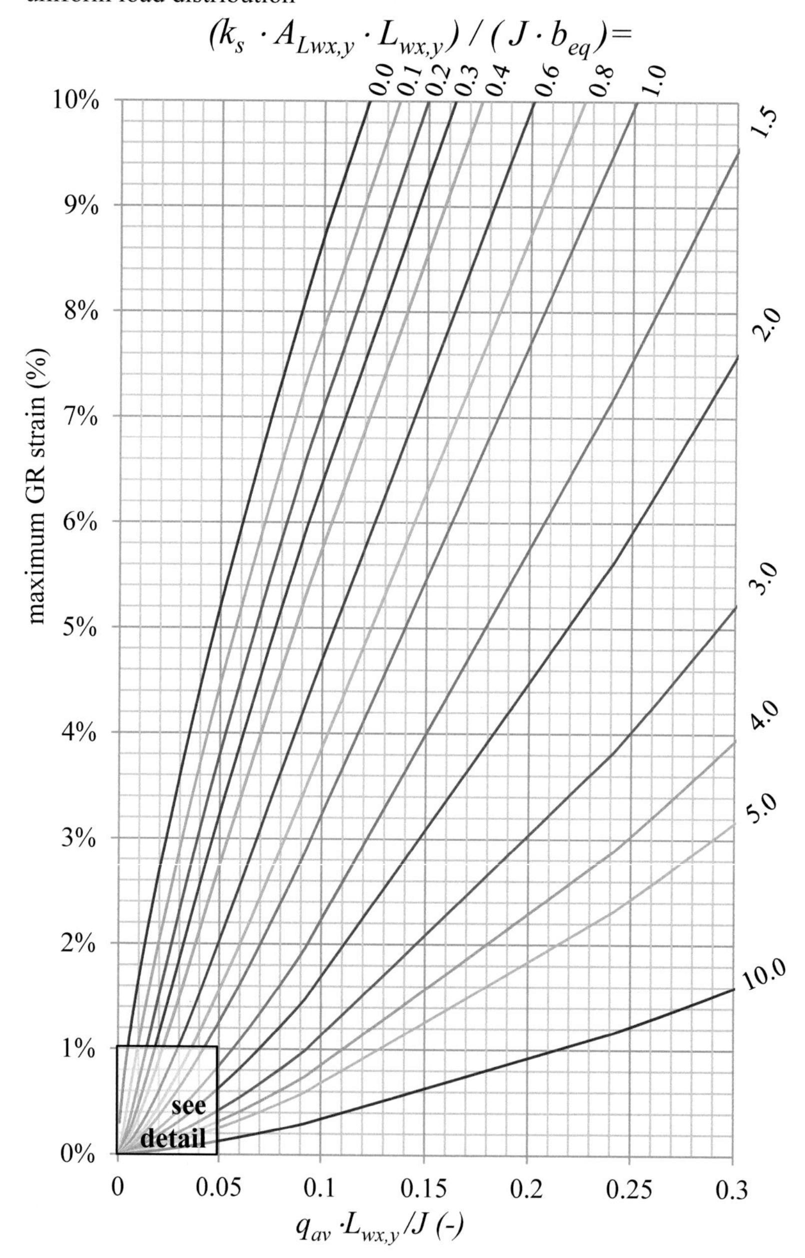

Fig. 4.9 Design chart: maximum GR strain ε_{max} in the reinforcement between the pile caps, uniform load distribution.

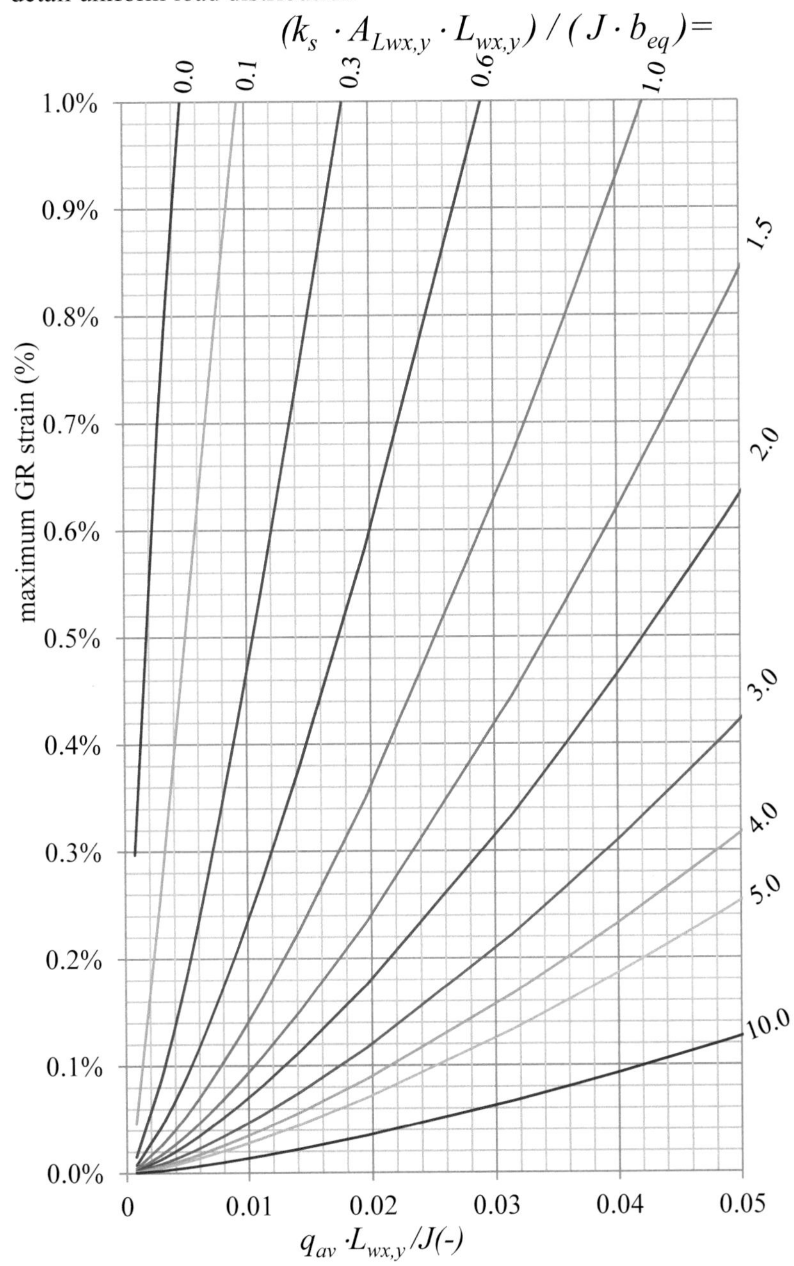

Fig. 4.10 Design chart: maximum GR strain ε_{max} in the reinforcement between the pile caps, detail for uniform load distribution.

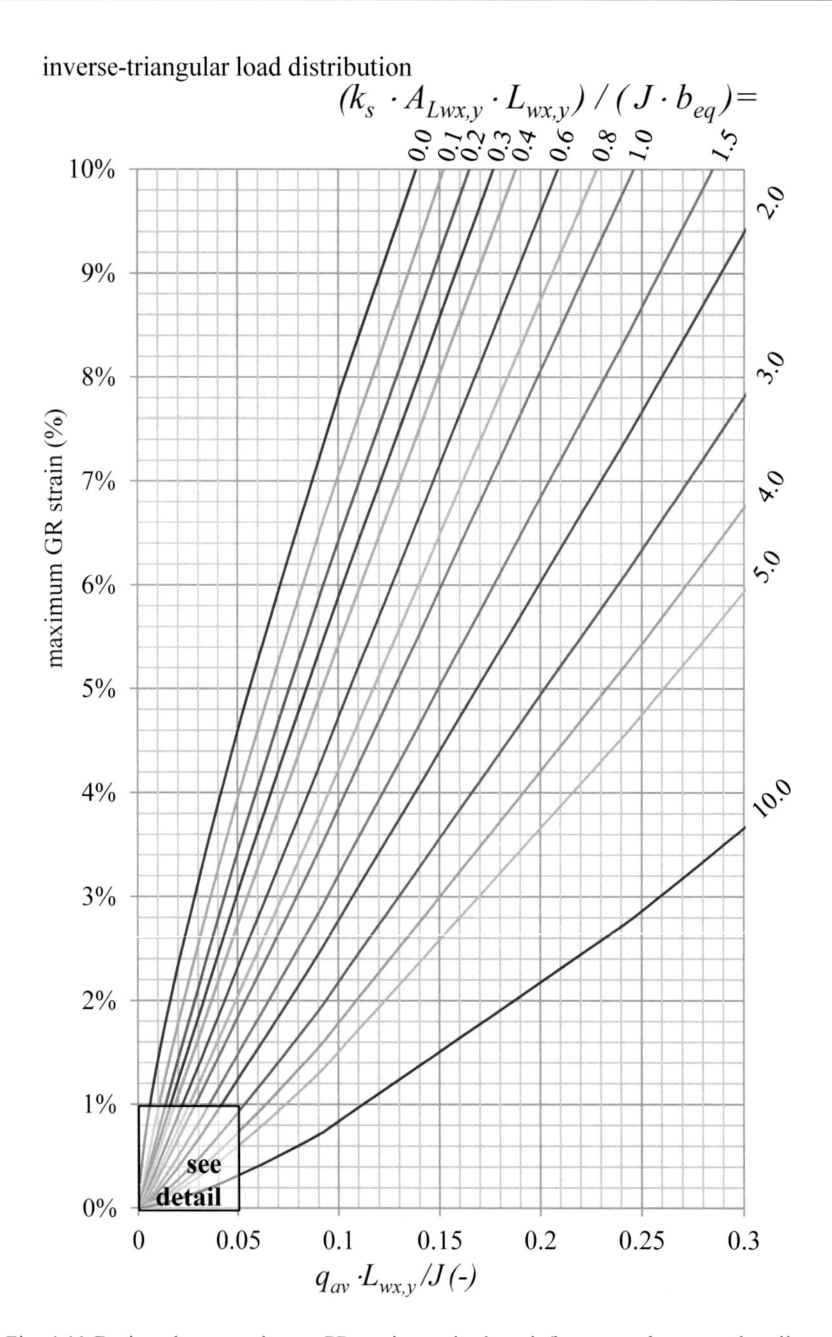

Fig. 4.11 Design chart: maximum GR strain ε_{max} in the reinforcement between the pile caps, inverse-triangular load distribution.

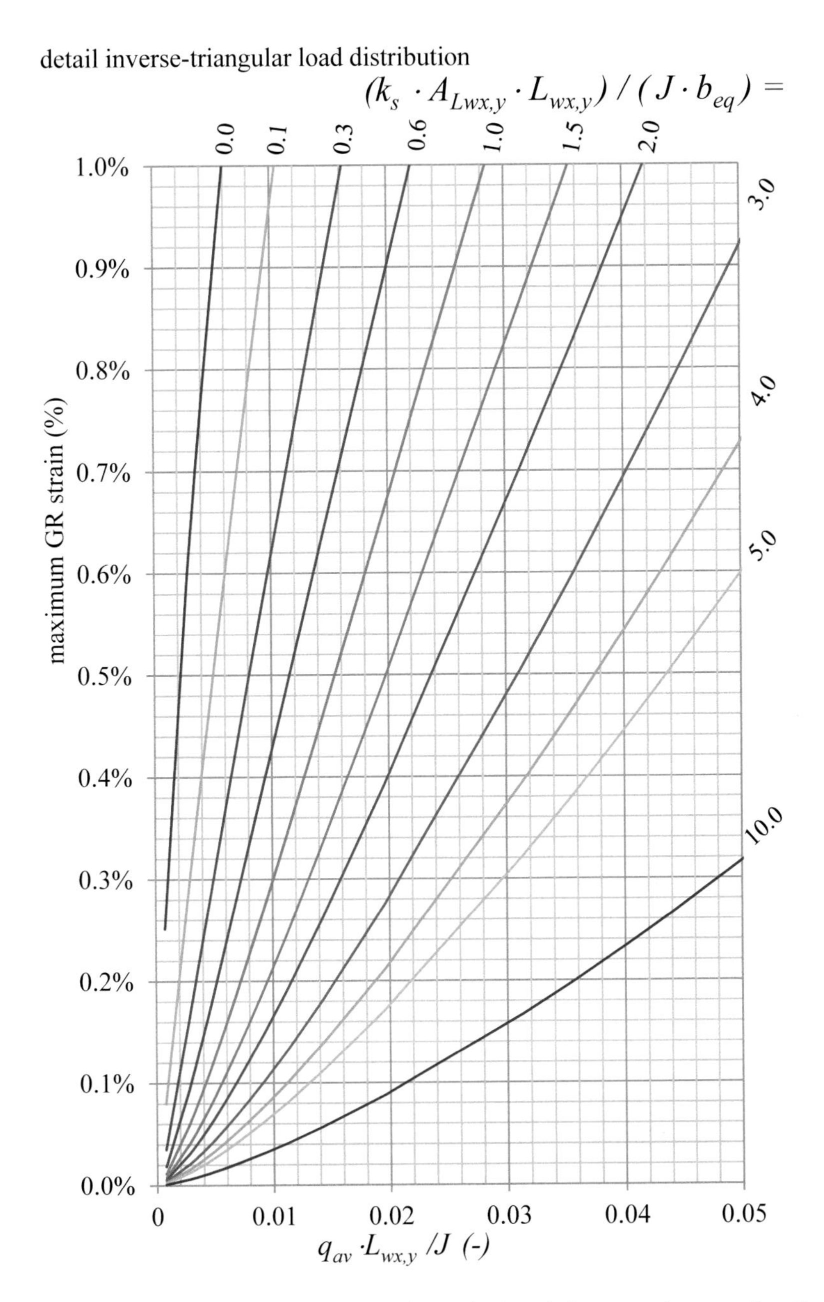

Fig. 4.12 Design chart: maximum GR strain ε_{max} in the reinforcement between the pile caps, detail for inverse-triangular load distribution.

4.3.3 *Tensile force due to horizontal load*

This Chapter describes how the additional GR tensile force due to lateral thrust is determined. The influence of braking forces and centrifugal forces in bends on strain and tensile force is described in Chapter 2.4. However, under the slope of the embankment, lateral thrust (lateral loading) arises locally as well as membrane forces (due to vertical loading); see Fig.4.13. The size of the lateral thrust depends on the embankment height, the fill frictional properties and the friction that is mobilised along the base of the embankment fill. A relatively simple approach to calculate the lateral thrust is given below. For a more detailed consideration, reference is made to EBGEO 2010.

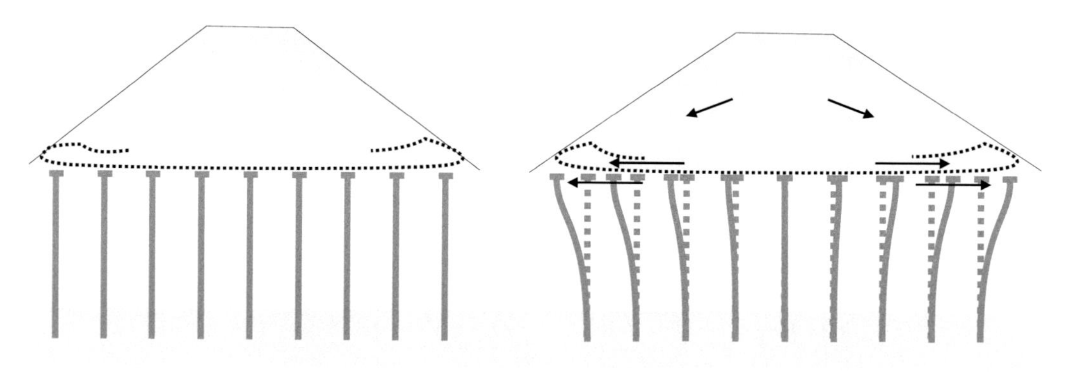

Fig. 4.13 GR tensile forces as a result of lateral thrust.

The lateral thrust is determined based on an assumed active soil pressure wedge that is built up from the top of the embankment to the reinforcement.

The lateral thrust can be calculated using (assuming that the distance between surcharge load and slope is zero):

$$T_{h,y} = F_{ea,h,a} = K_a \cdot \left[\tfrac{1}{2}\gamma\left(H - z\right) + p\right] \cdot \left(H - z\right) \quad (4.44)$$

where

$$K_a = \frac{1 - \sin\varphi}{1 + \sin\varphi} = \tan^2\left(45 - \frac{\varphi}{2}\right)$$

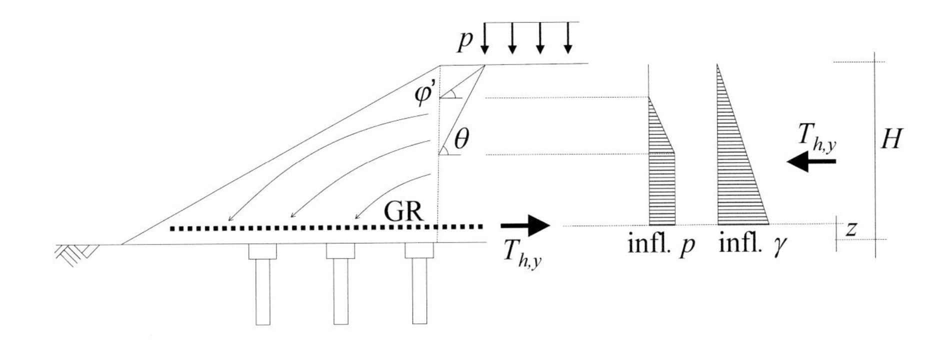

Fig. 4.14 Additional tensile force due to the lateral thrust generated in the geosynthetic reinforcement under the slope of the embankment.

4.3.4 *Total tensile force in geosynthetic reinforcement and checking*

The total GR tensile force in the geosynthetic reinforcement may be determined for the x and y directions from:

$$T_{s;x;d} = T_{v;x;d} \quad \text{and} \quad T_{s;y;d} = T_{v;y;d} + T_{h;y;d} \tag{4.45}$$

where:

$T_{s;x;d}$ $T_{s;y;d}$	kN/m	the design value of the total GR tensile force in the x and y directions respectively,
$T_{v;x,d}$ $T_{v;y;d}$	kN/m	the design value of the GR tensile force as a result of vertical loads acting in the GR strips in the x and y directions respectively in kN/m (calculated according to Chapter 4.3.2.6),
$T_{h;y;d}$	kN/m	the design value of the GR tensile force as a result of lateral outward thrust of the embankment fill in the y direction in kN/m (lateral spreading forces calculated according to Chapter 4.3.3),
x		direction parallel to the road axis,
y		direction perpendicular to the road axis.

The checking of the GR tensile force (ULS) against the design value of the GR tensile strength and the calculated GR strain (SLS) against the permissible strain is described in Chapter 2.10. The design values of the GR tensile forces are calculated in accordance with the safety approach described in Chapter 2.6 applying the associated partial factors listed in Chapter 2.7.

5 Calculation examples for the design of reinforced embankments

This chapter presents calculation examples for the design of the geosynthetic reinforcement for four different cases.

5.1 Requirements and initial details

Table 5.1 presents the requirements and initial details for design.

Table 5.1 Requirements and initial details[a]

		Parallel to road axis (x)	Transverse to road axis (y)
Maximum total GR strain at end of service life	%	5.0%	5.0%
Maximum GR strain during service life[b]	%	3.0%	3.0%
The design (ULS) value of the tensile force in the geosynthetic should be less than the design value of the tensile strength of the example geosynthetic[c] with a short-term tensile strength $T_{r,max}$ of:	kN/m	Case 1: 250 Case 2: 225	Case 1: 375 Case 2: 300
Traffic load		Two lanes and $N = 2$ million, so Table 2.3 applies. See Chapter 5.3.	
RC (reliability class)[d]		RC1 (roads)	

[a] Arbitrary example; see also the instructions in Chapter 4.2.2.

[b] A starting point in this calculation example is that the GR strain during the service life may not exceed 3.0%. To determine this strain during the service life, the SLS-GR strain after 1 year (end of construction phase) is compared with the SLS-GR strain after 120 years (end of service life). These two calculations differ in the following aspects:

- The subgrade reaction k_s (0 for 120 years, but > 0 for 1 year (k_s = 100 kN/m³ in this case)).
- The traffic load (construction traffic for one year that is less than the service life traffic for 120 years). For this calculation example, a construction traffic load of 0 kPa is applied because the chance exists that there are places where no or too little construction traffic has traversed over.
- The GR stiffness for 1 and 120 years respectively. The GR stiffness is determined using the isochronous curves in Chapter 5.5.

[c] At the start of a design process a suitable geosynthetic needs to be selected. In Chapter 5.5 this choice is used to determine the GR stiffness. In Chapter 5.6 it is checked whether the selected geosynthetic meets the deformation requirements and whether the strength is sufficient. If not, the calculations should be repeated with a different geosynthetic.

[d] For reliability class RC1 the partial factors for RC1 listed in Table 2.5 apply to the GR design.

5.2 Four cases: geometry, properties, load and approach

Table 5.2 presents the cases covered in this Chapter.

Table 5.2 Initial details for the four cases.

Case		1a	1b	1c	2	3	4[d]
GR load duration[a]		120 years		1 year (until handover)	120 years	N/A[b]	
		SLS	ULS	SLS			
s_x (m)	Centre-to-centre (ctc) pile spacing parallel to the road axis (longitudinal direction)	2.25					
s_y (m)	Centre-to-centre (ctc) pile spacing perpendicular to the road axis (transverse direction)	2.25			**2.00**	2.25	
H (m)	Total embankment height (including pavement construction)	3.50				1.60	0.60[d]
k_s (kN/m^3)	Subgrade reaction (subsoil support)	0 (no)		100 (yes)	0 (no)	N/A[b]	
p	Traffic	according to Chapter 2.3[c]		0 kPa	according to Chapter 2.3[c]		construction traffic[d]
	Braking load or centrifugal force?	braking load		no	no		no

[a] to determine the GR stiffness (Chapter 5.5)

[b] the calculation example only gives the step 1 results for cases 3 and 4 so these parameters are not applicable.

[c] with two lanes and N = 2 million; Table 2.3 applies; see details in Chapter 5.3.

[d] construction phase; no pavement construction; in this calculation example a construction traffic load of 8.6 kPa is applied.

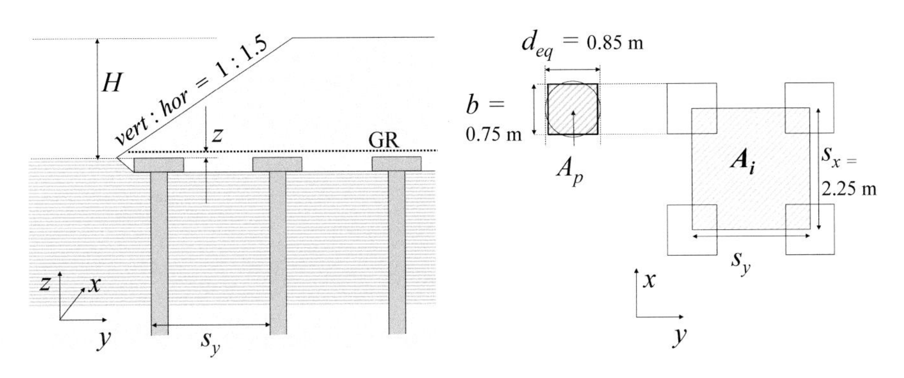

Fig. 5.1 Geometry of piled embankment

In addition, the following initial details are adopted for all cases:

Table 5.3 Initial details for all cases

$H_{asphalt}$	Thickness of asphalt layer	0.18 m (no pavement construction or road base in case 4)
$H_{road\ base}$	Thickness of road base	0.30 m (no pavement construction or road base in case 4)
b	Width of square pile cap	0.75 m
	Incline of embankment slope	Vertical : horizontal = 1 : 1.5
φ'_{cv}	Internal friction angle of granular fill	45°
γ	Unit weight of granular fill	19 kN/m³
J	Axial GR stiffness	According to isochronous curves, see Chapter 5.5.
z	Separation between top of pile cap and GR	0 m

This chapter first presents a calculation example for case 1. Cases 1a, 1b and 1c are all needed for this. The design calculations comprise the following steps:

- Calculation of design value of the traffic load (Chapter 5.3).
- Calculation of GR strains and GR tensile forces.
  - Calculation step 1 (arching, Chapter 5.4): division of the vertical load into two parts:
    - *A* in kN/pile (unit): the vertical load that goes directly to the pile caps and;
    - *B*+*C* in kN/pile: total vertical load on the GR between the piles (see Fig.2.1 for the explanation of load parts *A*, *B* and *C*).
  - Calculation step 2 (Chapter 5.5):
    - The GR stiffness from the isochronous curves;
    - The maximum GR strains ε_{max} and GR tensile forces $T_{v;d}$ that results from the vertical load *B*+*C* from calculation step 1;
    - The GR tensile force $T_{h;d}$ that results from the horizontal load (lateral thrust).
- A check of whether the geosynthetic assumed meets the requirements in Table 5.1 (Chapter 5.6).

Additionally, a construction phase may be included in the design in which the embankment height is less and a construction traffic load is applied but this is not included in this calculation example of case 1. After the traffic load calculations in Chapter 5.3, Chapters 5.4 to 5.6 present the calculations for case 1 in Table 5.2. The differences from the other cases are then determined in Chapters 5.7 to 5.9.

5.3 Traffic load and arching reduction coefficient

A uniformly distributed traffic load p is required for the calculations. This is determined using Table 2.3 (as there are two lanes). For this, the equivalent embankment height should be determined using equations (2.1) and (2.2). For cases 1 and 2, we find that:

$$h_{1,2;eq} = 0.9 \cdot h_{1,2} \cdot \sqrt[3]{\frac{E_{1,2}}{E_3}} \qquad (5.1)$$

using equation (2.2)

$$h_{1;eq} = 0.9 \cdot h_1 \cdot \sqrt[3]{\frac{E_1}{E_3}} = 0.9 \cdot 0.18 \cdot \sqrt[3]{\frac{8000}{200}} = 0.55 \text{ m}$$

$$h_{2;eq} = 0.9 \cdot h_2 \cdot \sqrt[3]{\frac{E_2}{E_3}} = 0.9 \cdot 0.30 \cdot \sqrt[3]{\frac{800}{200}} = 0.43 \text{ m}$$

$$h_3 = 3.50 - 0.18 - 0.30 = 3.02 \text{ m}$$

$$H_{eq} = h_{1;eq} + h_{2;eq} + h_3 = 0.55 + 0.43 + 3.02 = 4.00 \text{ m} \qquad (5.2)$$

using equation (2.1)

where

H_{eq}	= total equivalent layer thickness in m	
E_1	= dynamic stiffness modulus of asphalt	= 8000 MPa
E_2	= dynamic stiffness modulus of road base	= 800 MPa
E_3	= dynamic stiffness modulus of embankment fill	= 200 MPa

The SLS values for the traffic load p are now obtained from Table 2.3 (2 lanes).

- For case 1a, with H_{eq} = 4.00 m and $s_x \cdot s_y$ = 2.25·2.25 m^2, p = average (19.44 ; 18.90) = 19.2 kPa. For cases 1b and 1c, see Table 5.4.
- For case 2 (Table 5.2), the equivalent height of the embankment fill is also 4.00 m. For this case, s_x = 2.25 m and s_y = 2.00 m. To obtain the traffic load from Table 2.3, a pile arrangement of 2.00·2.00 m^2 is assumed. From Table 2.3 a traffic load p = 19.4 kPa is obtained. It should be noted that a slightly greater pile spacing yields a slightly higher design load.
- For case 3 (Table 5.2), the equivalent height of the embankment fill is 2.10 m. The pile arrangement is defined by $s_x = s_y$ = 2.25 m.

From this: p = average of (39.43 ; 36.77 ; 36.41 ; 34.22) = 36.7 kPa. It is observed that a reduced embankment height gives a slightly higher design load.

- For case 4 (Table 5.2), the equivalent height of the embankment fill is 0.60 m. The pile arrangement is defined by $s_x = s_y = 2.25$ m. From this: p = average of (77.53 ; 60.34) = 68.9 kPa. In this calculation example, a construction traffic load of 8.6 kPa has been assumed.

Table 5.4 presents the design values for the traffic load for each case.

Table 5.4 Traffic load in example calculations for four cases, determined using equations (2.1) and (2.2) and Table 2.3.

Case	**1a**	**1b**	**1c**	**2**	**3**	**4**
	SLS	ULS (RC1)	SLS			
Equivalent embankment height (m)	4.00				2.10	0.60
Design values for traffic load p (obtained from Table 2.3)	p = ***19.2 kPa***	$p = \gamma_{f;p} \cdot 19.2 = 1.05 \cdot 19.2 =$ **20.2 kPa**[b]	**0.0 kPa**	p = **19.4 kPa**	p = ***36.7* kPa**	p = ***8.6* kPa**
$p/\sigma_{v;tot}$	< 0.50				$0.60 > 0.50$	< 0.50
$H/(s_d - d)$	N/A: this value is needed to calculate the value of κ; not relevant here				$1.60/(\sqrt{(2.25^2 \cdot 2.25^2)} - 0.85) = 0.69$ for Fig. 5.2	N/A
Arching reduction factor κ obtained from Fig. 4.6 with frequency 1 Hz (for roads)	$\kappa = 1.0$				κ = **1.5** read off from Fig. 5.2	$\kappa = 1.0$
Equation (4.8) or (4.10) whichever is applicable	(4.8)[a]				(4.10)	(4.8)[a]

[a] Equation (4.8) is identical to equation (4.10) for $\kappa = 1.0$.

[b] For cases 1a and 1b, a braking force in the longitudinal direction is taken into account, as indicated in Table 5.2. This means that 20% extra traffic load is taken into account parallel to the road axis. So, in the longitudinal direction, the calculation is done with $1.2 \cdot 19.2 = 23.04$ kPa for case 1a and $1.2 \cdot 20.2 = 24.2$ kPa for case 1b.

In this calculation example, the traffic load is not increased by 20% in the transverse direction. This would have been necessary if centrifugal forces from the traffic had to be applied. This increase in traffic load is the result of braking forces only, as centrifugal forces have not been included when determining the value for the arching reduction factor κ.

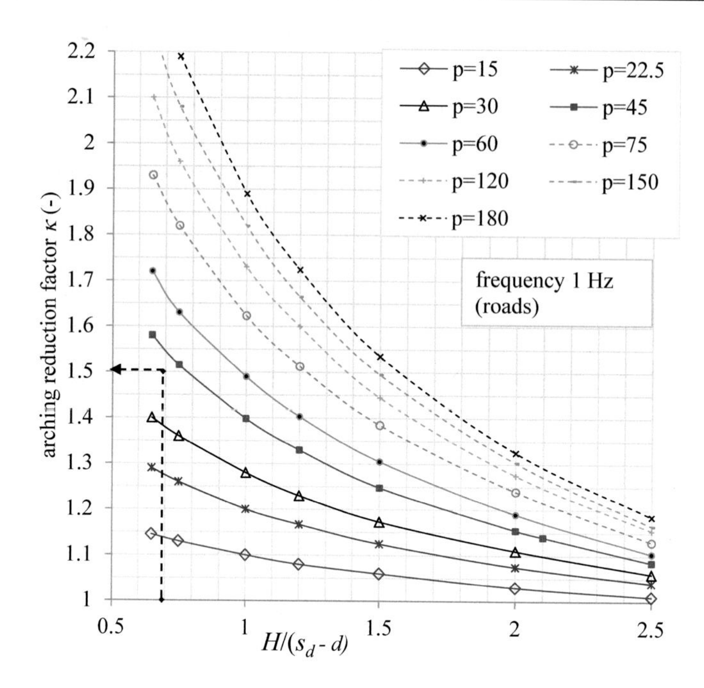

Fig.5.2 Determination of reduction factor κ using Fig. 4.6 for case 3, with $H/(s_d\text{-}d) = 0.69$ and $p = 36.7$ kPa; see Table 5.4.

5.4 Case 1, calculation step 1: load distribution

Determine vertical force on geosynthetic reinforcement between the pile caps for case 1:

Input parameters		Case 1a SLS After 120 years	Case 1b ULS After 120 years	Case 1c SLS On handover	
Diameter or equivalent diameter of pile cap	d_{eq}	0.85	0.85	0.85	m
Embankment height	H	3.50	3.50	3.50	m
Ctc pile spacing // road axis	s_x	2.25	2.25	2.25	m
Ctc pile spacing perpendicular to road axis	s_y	2.25	2.25	2.25	m
Vertical distance from average position of GR layers to top of pile cap	z	0.00	0.00	0.00	m
Unit weight of embankment	γ	19	20	19	kN/m^3
Traffic load	p	19.2	20.2	0.0	kPa
Friction angle	φ'	45.0	43.6	45.0	deg

Calculated parameters		Case 1a SLS After 120 years	Case 1b ULS After 120 years	Case 1c ULS On handover		Equation
Width or equivalent width of pile cap	b_{eq}	0.75	0.75	0.75	m	
Passive earth pressure coefficient	K_p	5.83	5.44	5.83	-	(4.15)
Height of 3D hemisphere	H_{g3D}	1.59	1.59	1.59	m	(4.13)
Width of GR square loaded by 3D hemi-spheres (see Fig. 5.3)	L_{3D}	1.50	1.50	1.50	m	(4.14)
Length of GR strips that are loaded by 2D arches (see Fig. 5.4)	L_{x2D}	1.50	1.50	1.50	m	(4.18)
	L_{y2D}	1.50	1.50	1.50	m	(4.18)
Calculated parameter	P_{x2D}	286.32	291.77	286.32	kPa/m^{Kp-1}	(4.21)
	P_{y2D}	286.32	291.77	286.32	kPa/m^{Kp-1}	(4.21)
Calculated parameter	$Q_{x2D} = Q_{y2D}$	28.93	31.61	28.93	kN/m^3	(4.21)
Calculated parameter	P_{3D}	2.16	3.00	2.16	kPa/m^{2Kp-2}	(4.12)
Calculated parameter	Q_{3D}	12.79	13.80	12.79	kN/m^3	(4.12)

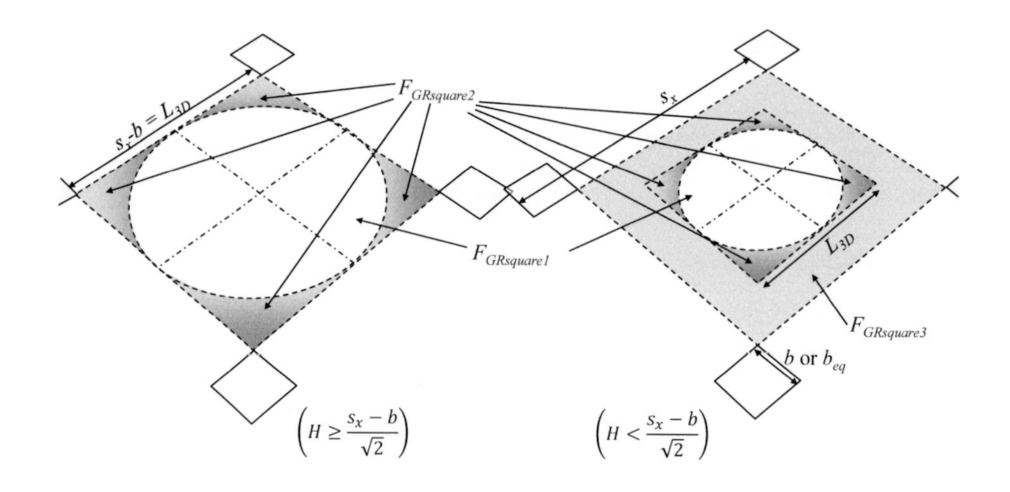

Fig. 5.3 Explanation of parameters (after Fig. 4.22 of Van Eekelen *et al.*, 2013 [21]).

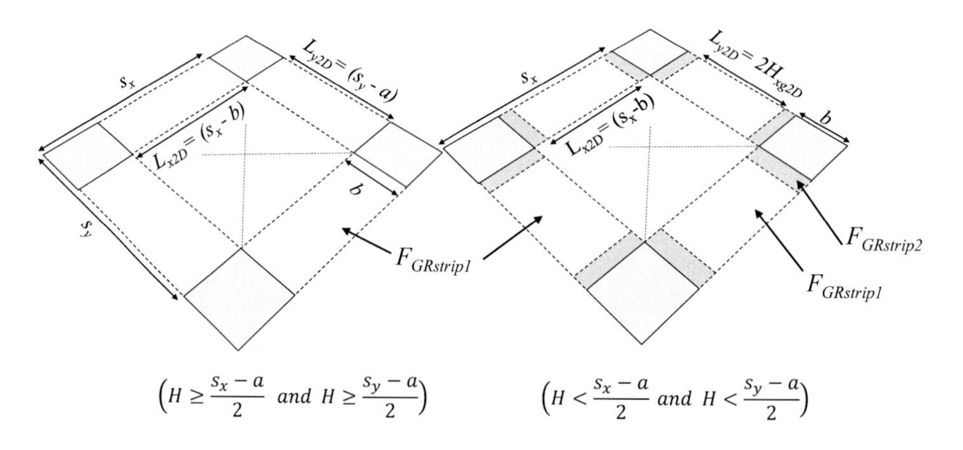

Fig. 5.4 Explanation of parameters (after Fig. 4.23 of Van Eekelen *et al.*, 2013 [21]).

Determination of the force exerted by the 3D hemispheres on the GR square (see Fig. 5.3; these calculations still neglect the surface load: $p = 0$):

		Case 1a SLS After 120 years	Case 1b ULS After 120 years	Case 1c SLS On handover		Equation
Force within the circle in the GR square (see Fig. 5.3)	$F_{GRsq1;p=0}$	11.34	12.27	11.34	kN/pile	(4.11)
Part 1 of the force on the area within the $L_{3D} \cdot L_{3D}$ square, but outside the circle (see Fig. 5.3)	${}_1F_{GRsq2}$	2.27	3.21	2.27	kN/pile	(4.11)
Part 2 of the force on the area within the $L_{3D} \cdot L_{3D}$ square, but outside the circle (see Fig. 5.3)	${}_2F_{GRsq2}$	20.67	22.30	20.67	kN/pile	(4.11)
Part 3 of the force on the area within the $L_{3D} \cdot L_{3D}$ square, but outside the circle (see Fig. 5.3)	${}_3F_{GRsq2}$	-1.97	-2.79	-1.97	kN/pile	(4.11)
Part 4 of the force on the area within the $L_{3D} \cdot L_{3D}$ square, but outside the circle (see Fig. 5.3)	${}_4F_{GRsq2}$	-15.45	-16.67	-15.45	kN/pile	(4.11)
Total force on the area within the $L_{3D} \cdot L_{3D}$ square, but outside the circle (see Fig. 5.3)	$F_{GRsq2;p=0}$	5.51	6.05	5.51	kN/pile	${}_1F_{GRsq2} + {}_2F_{GRsq2} + {}_3F_{GRsq2} + {}_4F_{GRsq2}$ (4.11)
Force on the area outside the $L_{3D} \cdot L_{3D}$ square, but within the GR square (see Fig. 5.3)	$F_{GRsq3;p=0}$	0.00	0.00	0.00	kN/pile	(4.16)
Total force on the GR square (see Fig. 5.3) with $p = 0$	$F_{GRsquare;p=0}$	16.86	18.32	16.86	kN/pile	(4.11)
Total force on the GR square (see Fig. 5.3) with $p > 0$, but still without braking load	$F_{GRsquare;p>0}$	21.73	23.60	16.86	kN/pile	(4.11)

Determination of the force that is transferred along the 3D hemispheres to the 2D arches. This load is applied as a surcharge load to the 2D arches:

		Case 1a SLS After 120 years	Case 1b SLS After 120 years	Case 1c SLS On handover		Equation
Transferred force (is calculated with $p = 0$)	$F_{transferred}$	132.77	139.18	132.77	kN/pile	(4.17)
Resulting surcharge load on the 2D arches (calculated with $p = 0$)	$p_{transferred}$	47.21	49.49	47.21	kPa	(4.17)

Determination of the load exerted by the 2D arches on the GR strips (see Fig. 5.4):

		Case 1a SLS After 120 years	Case 1b ULS After 120 years	Case 1c SLS On handover		Equation
Total force on the GR strips with $p = 0$ kPa	$F_{GRstrips;p=0}$	51.96	60.25	51.96	kN/pile	(4.20)
Total force on the GR strips with $p > 0$ kPa, but still without braking load	$F_{GRstrips;p>0}$	66.96	77.60	51.96	kN/pile	(4.20)

Determination of the load distribution:

		Case 1a SLS After 120 years	**Case 1b UGT After 120 years**	**Case 1c SLS On handover**		**Equation**
Total force on the GR with $p = 0$	$(B+C)^{stat}_{p=0}$	68.8	78.6	68.8	kN/pile	(4.8)
Force that goes directly via arching to the pile cap with $p = 0$	$A^{stat}_{p=0}$	267.9	275.8	267.9	kN/pile	(4.8)
Total force on the GR with $p > 0$, without braking load	$(B+C)^{stat}_{p>0}$	88.7	101.2	68.8	kN/pile	(4.9)
Average load on the GR strips with $p > 0$, without braking load	$q_{av;\, p>0}$	**39.4**	**45.0**	**30.6**	**kPa**	**(4.28)**
Force that goes directly via arching to the pile cap with $p > 0$	$A^{stat}_{p>0}$	**345.2**	**355.2**	**267.8**	**kN/pile**	**(4.9)**
Force that goes directly via arching to the pile cap with $p > 0$	$p_{A;\, p>0}$	**613.7**	**631.5**	**476.2**	**kPa**	$p_A = A / A_p$
Percentage of the total load that goes directly to the pile cap (via arching) with $p > 0$	$A\%;\, p>0$	79.6%	77.8%	79.6%	%	$A\% = A/((\gamma H+p)\cdot s_x \cdot s_y)$
Total force on the GR with $p > 0$, with braking load for end	$(B+C)^{stat}_{p>0}$	92.7	105.7	N/A	kN/pile	(4.9)
Average load on the GR strips with $p > 0$, with braking load	$q_{av;\, p>0}$	**41.2**	**47.0**	**N/A**	**kPa**	**(4.28)**

5.5 Case 1, calculation step 2: determine GR stiffness, calculation of GR strain and GR tensile force

The result from calculation step 1 is the vertical load on the geosynthetic: $B+C$. If $B+C$ is uniformly distributed over the geosynthetic (GR) strips, then the average load on the GR strips is q_{av}. Calculation step 2 uses q_{av} to calculate the strain and tensile force in the GR strips.

Two extra parameters are required for calculation step 2: the subgrade reaction k_s that is given in Table 5.2 and the GR stiffness J. The GR stiffness should be determined from the isochronous curves that present the time- and stress-dependent tensile stiffness (see Chapter 4.3.2.3).

The isochronous curves in Fig. 5.5 are for an example geosynthetic and serve to illustrate the calculation method only. For actual design it is necessary to use the isochronous curves from a real geosynthetic reinforcement material. These real curves used for design should be provided by the geosynthetic reinforcement supplier. The isochronous curves presented should have been determined according to the standards and guidelines in place.

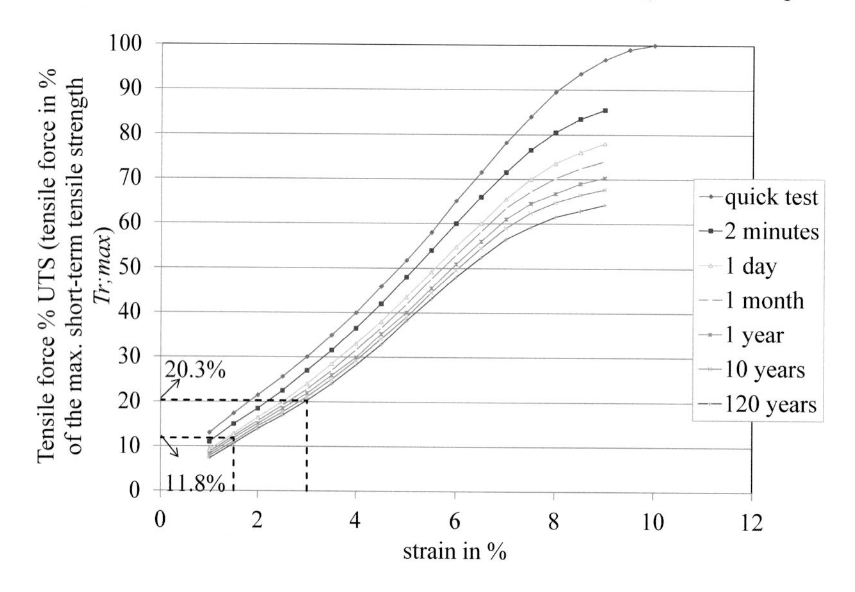

Fig. 5.5 Isochronous curves for the example geosynthetic reinforcement.

To determine the GR stiffness from these isochronous curves, a short-term GR tensile strength ($T_{r;max}$) should be chosen and the GR strain ε should be estimated. Afterwards, it is checked whether these assumptions were reasonable: the estimated GR strain should be roughly the same as the calculated $\varepsilon_{average}$. If this proves not to be the case, then the estimate of the strain ε and the resulting the GR stiffness and GR strength should be modified.

Table 5.5 Determining the GR stiffness for case 1 with the isochronous curves in Fig. 5.5.

	Parallel to the road axis (x)	Perpendicular to the road axis (y)
	$T_{r;x;max}$ = 250 kN/m (Table 5.1)	$T_{r;y;max}$ = 375 kN/m (Table 5.1)
Cases 1a and 1b	Degree of loading with ε = 3.0% and t = 120 years (obtained from Fig. 5.5): 20.3%	
Cases 1a and 1b	$J_{x;120yr}$ = 250·20.3% / 3.0% = 1692 kN/m	$J_{y;120yr}$ = 375·20.3% / 3.0% = 2538 kN/m
Case 1c	Degree of loading with ε = 1.5% and t = 1 year (obtained from Fig. 5.5): 11.8%	
Case 1c	$J_{x;120yr}$ = 250·11.8% / 1.5% = 1967 kN/m	$J_{y;120yr}$ = 375·11.8% / 1.5% = 2950 kN/m

The GR strain and GR tensile force are found by solving equation (4.37) numerically. This yields the solution for the unknown T_H that is shown in equations (4.31) and (4.32) for $z'(x)$ and in equation (4.34) for $T(x)$. Fig. 4.8 presents an example of a calculation chart that may be used for this purpose. An alternative approach to determine the GR strain is to use Fig. 4.9 to Fig. 4.12. In the rest of this calculation example both approaches are worked through.

In practice, cases 1a and 1b are used initially as these are often standard. In this example, we will start with case 1c because it is the only one applicable here that includes subsoil support and it is instructive to include it in a calculation example. To begin with, some intermediate results:

		Case 1c SLS On handover Parallel to road axis	**Case 1c SLS On handover Perpendicular to road axis**		**Equation/ Figure**
Average load on GR strips	q_{av}	30.6	30.6	kPa	Result from step 1
GR Stiffness	J	1967	2950	kN/m	Table 5.5
GR area associated with GR strip	A_L	2.25	2.25	m^2	(4.25) and (4.26)
Length of the GR strips	L_w	1.50	1.50	m	(4.27)
Subgrade reaction	k_s	100	100	kN/m^3	given
Modified subgrade reaction	K	200	200	kN/m^3	(4.29)
Term for Fig. 4.11 and Fig. 4.12	$q_{av}\cdot L_w/J$	0.023	0.016	-	
Term for Fig. 4.11 and Fig. 4.12	$k\cdot A_L\cdot L_w/(J\cdot b)$	0.229	0.153	-	

In a case where there is subsoil support the GR strain and GR tensile force should be determined using the two load distributions: the inverse-triangular and the uniform load distribution. This can be done in two ways:

1. Obtain ε_{max} from the charts in Fig. 4.9 to Fig. 4.12. For this, the last two terms in the above table are used. The maximum GR tensile force may then be found from: $T_{max} = \varepsilon_{max} \cdot J$. The values that are then determined are the same as (albeit possibly slightly less accurate than) the values that are calculated below.
2. Calculate ε_{max}. Here the maximum GR deflection can also be found as well as the departure angle of the geosynthetic reinforcement from the pile cap. This last value is required for the design of the pile cap; see Chapter 3.3. The calculations are firstly done for the inverse-triangular load distribution and then for the uniform load distribution.

Inverse-triangular load distribution		**Case 1c SLS On handover Parallel to road axis**	**Case 1c SLS On handover Perpendicular to road axis**		**Equation/ Figure**
Calculation parameter	M	0.57	0.63		(4.32)
Horizontal component of tensile force due to vertical load	T_H	28.07	33.91	kN/m	(4.37)
Calculation parameter	α	2.67	2.43	-	(4.33)
Max. GR strain (at edge of pile cap)	ε_{max} at $x,y=½(s_{x,y}-b)$	1.59%	1.25%	%	(4.38)[a]
Max. GR tensile force (at edge of pile cap, excluding lateral thrust)	T_{max} at $x,y=½(s_{x,y}-b)$	31.23	36.86	kN/m	(4.38)
Maximum GR deflection	z_{max}	0.08	0.07	m	(4.32)
Departure angle of GR from pile cap	$dz/dx,y$ at $x,y=½(s_{x,y}-b)$	26.6	23.6	deg	(4.32)
Average GR strain	$\varepsilon_{average}$	1.45%[b]	1.16%[b]	%	(4.37)

The average GR strain calculated in the SLS is compared to the maximum permitted strain:

Average strain · model factor (Table 2.5)	$\varepsilon_{average} \cdot \gamma_M$	**2.03%**	**1.63%**		**%**

[a] or read off from Fig. 4.11

[b] these values indeed lie in the order of 1.50%, as assumed in Fig. 5.5 and Table 5.5

Uniform load distribution		Case 1c SLS On handover Parallel to road axis	Case 1c SLS On handover Perpendicular to road axis		Equation/ Figure
Horizontal component of tensile force, due to vertical load	T_H	30.43	37.76	kN/m	(4.37)
Calculation parameter	α	2.56	2.30		(4.33)
Max. GR strain (at edge of pile cap)	ε_{max} at $x,y=½(s_{x,y}-b)$	1.65%	1.35%	%	(4.31)[a]
Max. GR tensile force (at edge of pile cap, excluding lateral thrust)	T_{max} at $x,y=½(s_{x,y}-b)$	32.50	39.77	kN/m	(4.31)
Maximum GR deflection	z_{max}	0.11	0.10	m	(4.31)
Departure angle of GR from pile cap	$dz/dx,y$ at $x,y= ½(s_{x,y}-b)$	21.1	18.7	m/m	(4.31)
Average GR strain	$\varepsilon_{average}$	1.57%[b]	1.30%[b]	%	(4.37)
The average strain calculated in the SLS is compared to the maximum permitted strain:					
Average strain · model factor (Table 2.5)	$\varepsilon_{average} \cdot \gamma_M$	2.20%	1.82%	%	

[a] or obtain from Fig. 4.9.

[b] these values are in the order of 1.50% as assumed in Fig. 5.5 and Table 5.5.

It emerges that the inverse-triangular load distribution yields smaller strains than the uniform load distribution and so the inverse-triangular load distribution should be maintained in this case. The average strain to be used, calculated with the inverse-triangular load distribution, is shown in bold on the previous page.

The consequent average strain from the inverse-triangular load distribution is shown in **bold** in the calculation above.

For cases 1a and 1b, there is no subsoil support. Here the inverse-triangular load distribution (equations (4.31)) should always be used. Below cases 1a and 1b are calculated:

		Cases 1a and 1 120 years SLS & ULS Parallel to road axis	Cases 1a and 1b 120 years SLS & ULS Perpendicular to road axis		Equation/ Figure
GR stiffness	J	1692	2538	kN/m	Tabel 5.5
GR area associated with GR strip	A_L	2.25	2.25	m^2	(4.25) and (4.26)
Length of the GR strips	L_w	1.50	1.50	m	(4.27)
Subgrade reaction	k_s	0	0	kN/m^3	given

For case 1a (120 years, SLS), it applies that:

		Case 1a SLS 120 years Parallel to road axis, without braking forces	case 1a SLS 120 years Parallel to road axis, with braking forces	Case 1a SLS 120 years Perpendicular to road axis		Equation/ Figure
Average load on GR strips	q_{av}	39.4	41.2	39.4	kPa	Result from step 1
Term for Fig. 4.11 and Fig. 4.12	$q_{av} \cdot L_w / J$	0.035	0.037	0.023	-	
Term for Fig. 4.11 and Fig. 4.12	$k \cdot A_L \cdot L_w / (J \cdot b)$	0.00	0.00	0.00	-	
Calculation parameter	M	1.00	1.00	1.00		(4.32)
Horizontal component of tensile force, due to vertical load	T_H	51.6	53.1	59.4	kN/m	(4.37)
Calculation parameter	α	0.0	0.0	0.0	-	(4.33)
Max. GR strain (at edge of pile cap)	ε_{max} at $x,y = \frac{1}{2}(s_{x,y} - b)$	3.52%	3.63%	2.62%	%	(4.38)[a]
Max. GR tensile force (at edge of pile cap, excluding lateral thrust)	T_{max} at $x,y = \frac{1}{2}(s_{x,y} - b)$	59.46	61.43	66.35	kN/m	(4.38)
Vertical component of maximum GR tensile force due to vertical load	$T_v = \sqrt{(T^2_{max} - T^2_H)}$	29.6	30.9	29.6	kN/m	
Part of the vertical load that is transferred via the GR to the piles (for design of pile cap)[c]	$B = T_v \cdot 4b$	88.7	92.7	88.7	kN/ pile	
Maximum GR deflection	z_{max}	0.14	N/A[e]	0.12	m	(4.32)
Departure angle of GR from pile cap	$dz/dx,y$ at $x,y = \frac{1}{2}(s_{x,y} - b)$	30.8	N/A[e]	27.3	deg	(4.32)
Average GR strain	$\varepsilon_{average}$	3.15%[b]	3.24%[b]	2.40%[b]	%	(4.37)

Model factor:

		Case 1a SLS 120 years Parallel to road axis, without braking forces	**Case 1a SLS 120 years Parallel to road axis, with braking forces**	**Case 1a SLS 120 years Perpendicular to road axis**		**Equation/ Figure**
Average strain · model factor (Table 2.5)[d]	$\varepsilon_{average} \cdot \gamma_M$	**4.40%**	**4.54%**	**3.36%**	**%**	
Max. GR tensile force including model factor	$T_{max} \cdot \gamma_M$	83.3	86.0	92.9	kN/m	
Part of the vertical load that is transferred via the GR to the piles (for design of pile cap)[c], including model factor	$B \cdot \gamma_M$	124.2	129.7	124.2	kN/ pile	

[a] or obtained from Fig. 4.11.

[b] these values lie in the order of 3.00%, as assumed in Fig. 5.5 and Table 5.5.

[c] these values are the same as $B+C$ because C is zero here as there is no subsoil support ($k = 0$ kN/m^3).

[d] the average GR strain calculated in the SLS is compared to the maximum permitted GR strain.

[e] the GR deflection and the departure angle is determined without consideration of braking or centrifugal loads, because these loads only cause a greater tensile force in the geosynthetic and not a greater GR deflection.

For case 1b (120 years, ULS):

		Case 1b ULS 120 years Parallel to road axis, without braking forces	**case 1b ULS 120 years Parallel to road axis, with braking forces**	**Case 1b ULS 120 years Perpendicular to road axis**		**Equation/ Figure**
Average load on GR strips	q_{av}	45.0	47.0	45.0	kPa	Result from step 1
Term for Fig. 4.11 and Fig. 4.12	$q_{av} \cdot L_w / J$	0.040	0.042	0.027	-	
Term for Fig. 4.11 and Fig. 4.12	$k \cdot A_L \cdot L_w / (J \cdot b)$	0.00	0.00	0.00	-	
Calculation parameter	M	1.00	1.00	1.00		(4.32)
Horizontal component of tensile force, due to vertical load	T_H	56.2	57.8	64.8	kN/m	(4.37)
Calculation parameter	α	0.0	0.0	0.0	-	(4.33)
Max. GR strain (at edge of pile cap)	ε_{max} at $x,y = ½(s_{x,y} - b)$	3.88%	4.00%	2.88%	%	(4.38)[a]
Max. GR tensile force (at edge of pile cap, excluding lateral thrust)	T_{max} at $x,y = ½(s_{x,y} - b)$	65.6	67.7	73.0	kN/m	(4.38)
Vertical component of maximum tensile force due to vertical load	$T_v = \sqrt{(T^2_{max} - T^2_H)}$	33.7	35.2	33.7	kN/m	
Part of the vertical load that is transferred via the geosynthetic to the piles (for design of pile cap)	$B = T_v \cdot 4b$	101.2	105.7	101.2	kN/pile	
Maximum GR deflection	z_{max}	0.15	N/A	0.13	m	(4.32)
Average GR strain	$\varepsilon_{average}$	3.44%	3.54%	2.62%	%	(4.37)
Maximum tensile force · model factor (Table 2.5)[b]	$T_{max} \cdot \gamma_M$	91.8	94.8	102.2	kN/m	
Horizontal component of tensile force due to vertical load with model factor	$T_H \cdot \gamma_M$	78.7	81.0	90.7	kN/m	

		Case 1b ULS 120 years Parallel to road axis, without braking forces	case 1b ULS 120 years Parallel to road axis, with braking forces	Case 1b ULS 120 years Perpendicular to road axis		Equation/ Figure
Part of the vertical load that is transferred via the GR to the piles (for design of pile cap), including model factor	$B \cdot \gamma_M$	141.7	148.0	141.7	kN/pile	
Active earth pressure coefficient	K_a			0.184	-	(4.44)
Lateral thrust (spreading force, only in the transverse direction, in the vicinity of a slope)	T_h	-		35.5	kN/m	(4.44)
Total tensile force	$T_{max,tot}$	**91.8**	**94.8**	**137.7**	**kN/m**	(4.45)

[a] or obtained from Fig. 4.11

[b] The maximum GR tensile force calculated in the ULS is compared with the design value of the GR strength.

5.6 Check of case 1

It should be checked whether the requirements on GR strains and strength contained in Table 5.1 are met.

Table 5.6 Check on GR strains (deformations).

		Parallel to the road axis	Perpendicular to the road axis
Calculated average strain $\varepsilon_{average}$	After 120 years SLS (case 1a)	4.54%[a]	3.36%
	On handover SLS (case 1c)	2.03%	1.63%
	Creep (cases 1a – 1c)	2.51%	1.73%
Maximum permissible strain	Total strain at end of service life	< 5.0%	< 5.0%
	Strain during service life	< 3.0%	< 3.0%
Check		4.54% < 5.0% and 2.51% < 3.0% → pass	3.36% < 5.0% and 1.73% < 3.0% → pass

[a] including braking load.

Table 5.7 Check on GR strength.

		Parallel to the road axis	Perpendicular to the road axis
Calculated tensile force	After 120 years ULS (case 1b)	94.8 kN/m[a]	137.7 kN/m
Strength (see also Chapter 2.10.1)	Characteristic value of tensile strength, short-term $T_{r;st;k}$	250.0 kN/m	375.0 kN/m
	Characteristic value of tensile strength, long-term $T_{r;lt;k} = T_{r;st;k}/(A_1 \cdot A_2 \cdot A_3 \cdot A_4 \cdot A_5)$[b]	128.0 kN/m	192.0 kN/m
	Design value of tensile strength, long-term $T_{r;lt;d} = T_{r;lt;k}/\gamma_{m;T}$ (according to Table 2.5 for RC1: $\gamma_{m;T} = 1.30$)	98.4 kN/m	147.7 kN/m
Check		94.8 kN/m < 98.4 kN/m → pass	137.7 kN/m < 147.7 kN/m → pass

[a] including braking load.

[b] A_1, A_2, A_3, A_4 and A_5 are reduction factors that should be defined according to the relevant standards in place. In this calculation example, the following values are used:
creep: $A_1 = 1.48$; installation damage: $A_2 = 1.20$; overlaps: $A_3 = 1.0$ (no reduction in the strength of overlaps as sufficiently broad overlaps are used); chemical effects $A_4 = 1.10$; dynamic load $A_5 = 1.0$.

5.7 Case 2: unequal centre-to-centre pile spacing longitudinally and transversely

Only the SLS for 120 years is presented here. The results for step 1 of case 2 are:

Diameter or equivalent diameter of pile cap	d_{eq}	0.85	m	
Embankment height	H	3.50	m	
Ctc pile spacing // road axis	s_x	2.25	m	
Ctc pile spacing perpendicular to road axis	s_y	2.00	m	
Vertical distance from average position of GR layers to top of pile cap	z	0.00	m	
Unit weight of embankment	γ	19	kN/m^3	
Traffic load	p	19.4	kPa	
Friction angle	φ'	45.0	deg	
Width or equivalent width of pile cap	b_{eq}	0.75	m	
Passive earth pressure coefficient	K_p	5.83	-	(4.15)
Height of 3D hemisphere	H_{g3D}	1.51	m	(4.13)
Width of GR square loaded by 3D hemispheres (see Fig. 5.3)	L_{3D}	1.38	m	(4.14)
Length of GR strips that are loaded by 2D arches (see Fig. 5.4)	L_{x2D}	1.50	m	(4.18)
	L_{y2D}	1.25	m	(4.18)
Calculated parameter, longitudinal and transverse	P_{x2D}	273.4	kPa/m^{Kp-1}	(4.21)
	P_{y2D}	500.3	kPa/m^{Kp-1}	(4.21)
Calculated parameter	$Q_{x2D} = Q_{y2D}$	28.93	kN/m^3	(4.21)
Calculated parameter	P_{3D}	3.89	kPa/m^{2Kp-2}	(4.12)
Calculated parameter	Q_{3D}	12.79	kN/m^3	(4.12)
Force within the circle in the GR square	$F_{GRsq1;p=0}$	8.84	kN/pile	(4.11)
Part 1 of the force on the area within the $L_{3D}\cdot L_{3D}$ square, but outside the circle	${}_1F_{GRsq2}$	1.56	kN/pile	(4.11)
Part 2 of the force on the area within the $L_{3D}\cdot L_{3D}$ square, but outside the circle	${}_2F_{GRsq2}$	16.12	kN/pile	(4.11)
Part 3 of the force on the area within the $L_{3D}\cdot L_{3D}$ square, but outside the circle	${}_3F_{GRsq2}$	-1.35	kN/pile	(4.11)
Part 4 of the force on the area within the $L_{3D}\cdot L_{3D}$ square, but outside the circle	${}_4F_{GRsq2}$	-12.05	kN/pile	(4.11)

Total force on the area within the $L_{3D}{\cdot}L_{3D}$ square, but outside the circle (see Fig. 5.3)	$F_{GRsq2;p=0}$	4.27	kN/pile	$_1F_{GRsq2}+{}_2F_{GRsq2}+{}_3F_{GRsq2}+{}_4F_{GRsq2}$ (4.11)
Force on the area outside the $L_{3D}{\cdot}L_{3D}$ square, but within the GR square (see Fig. 5.3)	$F_{GRsq3;p=0}$	0.00	kN/pile	(4.16)
Total force on the GR square (see Fig. 5.3)	$F_{GRsquare;p=0}$	13.11	kN/pile	(4.11)
Transferred force	$F_{transferred}$	111.57	kN/pile	(4.17)
Resulting surcharge load on the 2D arches	$p_{transferred}$	42.50	kPa	(4.17)
Total force on the GR strips	$F'_{GRstrips;p=0}$	41.95	kN/pile	(4.20)
Total force on the GR with $p = 0$	$(B+C)^{stat}_{p=0}$	55.06	kN/pile	(4.8)
Force that goes directly via arching to the pile cap with $p = 0$	$A^{stat}_{p=0}$	244.19	kN/pile	(4.8)
Total force on the GR with $p > 0$	$(B+C)^{stat}_{p>0}$	71.13	kN/pile	(4.9)
Average load on the GR strips with $p > 0$	q_{av}	**34.49**	**kPa**	**(4.28)**
Load that goes directly via arching to the pile cap with $p > 0$	$A^{stat}_{p>0}$	315.42	kN/pile	(4.9)
Load that goes directly via arching to the pile cap with $p > 0$	p_A	560.75	kPa	$p_A = A / A_p$
Percentage of the total load that goes directly to the pile cap (via arching) with $p > 0$	$A\%$	81.6%	%	$A\% = A/((\gamma H+p)\cdot s_x\cdot s_y)$

Table 5.8 Results of the determination of the GR stiffness in case 2 using the isochronous curves in Fig. 5.5

Parallel to the road axis (x)	**Perpendicular to the road axis (y)**
$T_{r;x;max} = 225$ kN/m	$T_{r;y;max} = 300$ kN/m
Degree of loading using $\varepsilon_x = 3.0\%$ and $t = 120$ years (obtained from Fig. 5.5): 20.3%	Degree of loading using $\varepsilon_y = 2.0\%$ and $t = 120$ years (obtained from Fig. 5.5): 13.9%
$J_{x;120yr} = 225{\cdot}20.3\% / 3.0\% = 1523$ kN/m	$J_{y;120yr} = 300{\cdot}13.9\% / 2.0\% = 2085$ kN/m

The results from step 2 for case 2 (SLS, 120 years) are:

		Case 2 SLS 120 years Parallel to road axis (x)	**Case 2 SLS 120 years Perpendicular to road axis (y)**		**Equation/ Figure**
GR stiffness	J	1523	2085	kN/m	Table 5.5
GR area associated with GR strip	A_L	1.99	1.95	m^2	(4.25) and (4.26)
Length of the GR strips	L_w	1.50	1.25	m	(4.27)
Subgrade reaction	k_s	0	0	kN/m^3	given
Average load on GR strips	q_{av}	34.49	34.49	kPa	Result from step 1
Term for Fig. 4.11 and Fig. 4.12	$q_{av} \cdot L_w / J$	0.0340	0.0207	-	
Term for Fig. 4.11 and Fig. 4.12	$k \cdot A_L \cdot L_w / (J \cdot b)$	0.00	0.00	-	
Calculation parameter	M	1.00	1.00	-	(4.32)
Horizontal component of tensile force due to vertical load	T_H	45.6	45.2	kN/m	(4.37)
Calculation parameter	α	0.0	0.0	-	(4.33)
Max. GR strain (at edge of pile cap)	ε_{max} at $x,y = ½(s_{x,y} - b)$	3.44%	2.40%	%	(4.38)[a]
Max. tensile force (at edge of pile cap, excluding lateral thrust)	T_{max} at $x,y = ½(s_{x,y} - b)$	52.4	50.0	kN/m	(4.38)
Maximum GR deflection	z_{max}	0.15	0.10	m	(4.32)
Departure angle of GR from pile cap	$dz/dx,y$ at $x,y = ½(s_{x,y} - b)$	29.6	25.5	deg	(4.32)
Average GR strain	$\varepsilon_{average}$	3.1%[b]	2.2%[c]	%	(4.37)
The average strain calculated in the SLS is compared with the maximum permitted strain.					
Average strain · model factor (Table 2.5)	$\varepsilon_{average} \cdot \gamma_M$	**4.3%**	**3.1%**	**%**	

[a] or obtained from Fig. 4.11

[b] this value lies in the order of 3.00% as assumed in Fig. 5.5 and Table 5.8

[c] this value lies in the order of 2.00% as assumed in Fig. 5.5 and Table 5.8

5.8 Case 3: shallow embankment ($\kappa > 1.0$)

Only the SLS is presented here. The results for step 1 of case 3 are:

Diameter or equivalent diameter of pile cap	d_{eq}	0.85	m	
Embankment height	H	1.60	m	
Ctc pile spacing // road axis	s_x	2.25	m	
Ctc pile spacing perpendicular to road axis	s_y	2.25	m	
vertical distance from average position of GR layers to top of pile cap	z	0.00	m	
Unit weight of embankment	γ	19	kN/m^3	
Traffic load	p	36.7	kPa	
Friction angle	φ'	45.0	deg	
Width or equivalent width of pile cap	b_{eq}	0.75	m	
Passive earth pressure coefficient	K_p	5.83	-	(4.15)
Height of 3D hemisphere	H_{g3D}	1.59	m	(4.13)
Width of GR square loaded by 3D hemispheres (see Fig. 5.3)	L_{3D}	1.50	m	(4.14)
Length of GR strips that are loaded by 2D arches (see Fig. 5.4)	L_{x2D}	1.50	m	(4.18)
	L_{y2D}	1.50	m	(4.18)
Calculated parameter, longitudinal and transverse	P_{x2D}	72.28	kPa/m^{Kp-1}	(4.21)
	P_{y2D}	72.28	kPa/m^{Kp-1}	(4.21)
Calculated parameter	$Q_{x2D} = Q_{y2D}$	28.93	kN/m^3	(4.21)
Calculated parameter	P_{3D}	-0.22	kPa/m^{2Kp-2}	(4.12)
Calculated parameter	Q_{3D}	12.79	kN/m^3	(4.12)
Force within the circle in the GR square	$F_{GRsq1;p=0}$	11.30	kN/pile	(4.11)
Part 1 of the force on the area within the $L_{3D} \cdot L_{3D}$ square, but outside the circle	${}_1F_{GRsq2}$	-0.23	kN/pile	(4.11)
Part 2 of the force on the area within the $L_{3D} \cdot L_{3D}$ square, but outside the circle	${}_2F_{GRsq2}$	20.67	kN/pile	(4.11)
Part 3 of the force on the area within the $L_{3D} \cdot L_{3D}$ square, but outside the circle	${}_3F_{GRsq2}$	0.20	kN/pile	(4.11)
Part 4 of the force on the area within the $L_{3D} \cdot L_{3D}$ square, but outside the circle	${}_4F_{GRsq2}$	-15.45	kN/pile	(4.11)
Total force on the area within the $L_{3D} \cdot L_{3D}$ square, but outside the circle (see Fig. 5.3)	$F_{GRsq2;p=0}$	5.19	kN/pile	${}_1F_{GRsq2}+ {}_2F_{GRsq2}+ {}_3F_{GRsq2}+ {}_4F_{GRsq2}$ (4.11)

Force on the area outside the $L_{3D} \cdot L_{3D}$ square, but within the GR square (see Fig. 5.3)	$F_{GRsq3;p=0}$	0.00	kN/pile	(4.16)
Total force on the GR square (see Fig. 5.3)	$F_{GRsquare;p=0}$	16.49	kN/pile	(4.11)
Transferred force	$F_{transferred}$	51.91	kN/pile	(4.17)
Resulting surcharge load on the 2D arches	$p_{transferred}$	18.46	kPa	(4.17)
Total force on the GR strips	$F_{GRstrips;p=0}$	31.36	kN/pile	(4.20)
Total force on the GR with $p = 0$	$(B+C)_{p=0}^{stat}$	47.6	kN/pile	(4.8)
Total static force on GR with $p > 0$	$(B+C)_{p>0}^{stat}$	105.6	kN/pile	(4.9)
Total cyclic force on GR with p > 0 with $\kappa = 1.50$ (see Table 5.4)	$(B+C)_{p>0}^{cycl}$	183.6	kN/pile	(4.10)
Average load on the GR strips with $p > 0$	$q_{av\ (cycl)}$	**81.6**	**kPa**	**(4.28)**

5.9 Case 4: very shallow embankment (construction phase, partial arching)

Only the SLS is presented here. The results for step 1 of case 3 are:

Diameter or equivalent diameter of pile cap	d_{eq}	0.85	m	
Embankment height	H	0.60	m	
Ctc pile spacing // road axis	s_x	2.25	m	
Ctc pile spacing perpendicular to road axis	s_y	2.25	m	
Vertical distance from average position of GR layers to top of pile cap	z	0.00	m	
Unit weight of embankment	γ	19	kN/m³	
Unit weight of embankment	p	8.6	kPa	
Traffic load	φ'	45.0	deg	
Width or equivalent width of pile cap	b_{eq}	0.75	m	
Passive earth pressure coefficient	K_p	5.83	-	(4.15)
Height of 3D hemisphere	H_{g3D}	0.60	m	(4.13)
Width of GR square loaded by 3D hemispheres (see Fig. 5.3)	L_{3D}	0.85	m	(4.14)
Length of GR strips that are loaded by 2D arches (see Fig. 5.4)	L_{x2D}	1.20	m	(4.18)
	L_{y2D}	1.20	m	(4.18)
Calculated parameter, longitudinal and transverse	P_{x2D}	-46.44	kPa/m^{Kp-1}	(4.21)
	P_{y2D}	-46.44	kPa/m^{Kp-1}	(4.21)
Calculated parameter	$Q_{x2D} = Q_{y2D}$	28.93	kN/m³	(4.21)
Calculated parameter	P_{3D}	-1065.26	kPa/m^{2Kp-2}	(4.12)
Calculated parameter	Q_{3D}	12.79	kN/m³	(4.12)
Force within the circle in the GR square	$F_{GRsq1;p=0}$	2.02	kN/pile	(4.11)
Part 1 of the force on the area within the $L_{3D} \cdot L_{3D}$ square, but outside the circle	${}_1F_{GRsq2}$	-1.46	kN/pile	(4.11)
Part 2 of the force on the area within the $L_{3D} \cdot L_{3D}$ square, but outside the circle	${}_2F_{GRsq2}$	3.74	kN/pile	(4.11)
Part 3 of the force on the area within the $L_{3D} \cdot L_{3D}$ square, but outside the circle	${}_3F_{GRsq2}$	1.27	kN/pile	(4.11)
Part 4 of the force on the area within the $L_{3D} \cdot L_{3D}$ square, but outside the circle	${}_4F_{GRsq2}$	-2.80	kN/pile	(4.11)

Total force on the area within the $L_{3D} \cdot L_{3D}$ square, but outside the circle (see Fig. 5.3)	$F_{GRsq2;p=0}$	0.75	kN/pile	$_1F_{GRsq2}$ + $_2F_{GRsq2}$ + $_3F_{GRsq2}$ + $_4F_{GRsq2}$ (4.11)
Force on the area outside the $L_{3D} \cdot L_{3D}$ square, but within the GR square (see Fig. 5.3)	$F_{GRsq3;p=0}$	17.44	kN/pile	(4.16)
Total force on the GR square (see Fig. 5.3)	$F_{GRsquare;p=0}$	20.21	kN/pile	(4.11)
Transferred force	$F_{transferred}$	5.44	kN/pile	(4.17)
Resulting surcharge load on the 2D arches	$p_{transferred}$	2.30	kPa	(4.17)
Force on x GR strip outside 2D arches	$F_{xGRstr2p=0}$	2.57	kN/pile	(4.21)
Force on y GR strip outside 2D arches	$F_{yGRstr2p=0}$	2.57	kN/pile	(4.21)
Total force on the GR strips	$F_{GRstrips;p=0}$	19.53	kN/pile	(4.20)
Total force on the GR with $p = 0$	$(B+C)^{stat}_{p=0}$	39.75	kN/pile	(4.8)
Total static force on GR with $p > 0$	$(B+C)^{stat}_{p>0}$	69.7	kN/pile	(4.9)

6 Numerical modelling

6.1 Introduction

Numerical modelling may be desirable because analytical methods cannot determine horizontal deformations of the basal reinforced piled embankment, deformations in the vicinity of the embankment, and lateral pile loads and moments. Insight into the deformations and pile moments is important in connection with deformation requirements and the dimensioning of the piles. The amount of reinforcement in the piles forms a factor that should not be underestimated in their cost and produceability and thus in the basic feasibility of the piled embankment.

This chapter considers the numerical modelling of the piled embankment.

It should be stated emphatically that the designer who wishes to generate a numerical model should be fully aware of the extent and limitations of the program for modelling the structure or components of the structure. Further, reference should be made to Chapter 4.3, where it is stated that the strength calculation of the geosynthetic reinforcement for the ultimate limit state should be conducted analytically.

6.2 Purpose of numerical modelling

For design, numerical modelling of piled embankments is necessary to analyse:

- Bending moments and transverse shear forces in the piles under the influence of the weight of the embankment and surcharge loads;
- Horizontal deformations of the piles, the reinforced embankment and the embankment;
- Deformations and other influences in the vicinity of the piled embankment;
- Complex geometry such as the influence of an existing soil body or one to be constructed near the piled embankment;
- The consequences of horizontal traffic loads such as braking and acceleration forces and centrifugal forces in bends.

6.3 2D modelling

To begin with, the choice should be made between a 2D and a 3D approach for the numerical model. In current (2016) design practice, 3D calculation is not commonly used due to the substantially greater modelling and calculation time demanded. It is therefore common to model the structure two-dimensionally, particularly for linear and plane-strain geometries, where the choice of 2D modelling seems sufficiently accurate. Separate calculations will often be needed for the transverse and longitudinal cross-sections.

When modelling a basal reinforced piled embankment using a 2D model, the following aspects deserve particular attention:

- the stiffness of the pile;
- the behaviour of the soil between the piles (the wedge mechanism);
- the pile's subsidence behaviour, particularly when modelled as a 'wall' element, compared to that of an actual pile;
- to determine the bending moments and transverse shear forces in the pile the vector sum of the loads found in the transverse and longitudinal directions should be calculated. The normal forces found should be broadly equal to each other. To check the normal stresses in the pile, the lower normal force calculated is used.

6.4 Numerical model

When setting up a numerical model, the designer should make a number of choices that depend on the program being used, the purpose for which the model is being set up, insitu conditions such as the soil profile, the embankment structure and its complexity, and the phasing of the construction. It is impossible to account for all aspects in this context. The model should however specifically include or meet the following conditions:

- constructional phasing that reproduces the actual phasing;
- an 'air gap' between the piles immediately below the geosynthetic reinforcement if support pressure by the subsoil is not guaranteed;
- incorporation of second order effects (geometrical non-linearity).

6.5 Deformation calculation, SLS

For checking the deformations at the serviceability limit state using a numerical model, characteristic values of the parameters should be used. Thus, an upper limit for the deformation is calculated. In special cases, it may be beneficial to conduct a calculation using average values of the parameters for the verification of deformation measurements on structures during the construction phase or the service life.

6.6 Calculation design values of bending moment and transverse shear force in the piles, ULS

With numerical models, characteristic values of the parameters are generally used because the model then yields realistic results and is more stable. For the dimensioning of the reinforcement in the piles however, the design value of the pile moment at the ultimate limit state is required. The moments in the piles may be determined in two ways:

A. Calculation using design values
In this approach, the calculations are conducted using:

- Design values for both the strength of the soil and the loads.
- Low characteristic values (lower boundary of the 95% confidence interval) for the stiffness of the soil and structural elements (geosynthetics, piles).

The disadvantage of this approach is that (due to adopting design values for soil strengths and loads) the displacements in each construction phase are overestimated and as such, the calculated forces and moments are also overestimated. Thus adopting this design approach may not result in the most optimal design.

B. Calculation using representative/characteristic values
In this approach, the calculations are done using:

- Characteristic values for both the strength of the soil and the loads. The most unfavourable combination of the lower- and higher boundary values of the 95% confidence interval of these parameters should be used.
- Low characteristic values for the soil stiffness and the structural elements (geosynthetics, piles).
- The required phase should then be analysed using the design values for both the strength parameters of the soil and the loads.

The advantage of this calculation approach is that a better insight is gained into the resulting deformations, hence the resulting forces and moments are less conservative.

In addition to both approaches A and B, a calculation should be done using characteristic values (the most unfavourable combination of the lower- and higher boundary values of the 95% confidence interval) for all parameters. The resulting characteristic shear forces and bending moments from the relevant construction phase should be multiplied by a factor of 1.2 and compared with the design value of the forces and moments calculated with approach A or B. The normative forces and moments should then be used in the structural evaluation of the pile.

The partial factors to be employed in order to obtain the design values of the parameters for input into the numerical model to determine the moments and transverse forces in the piles are summarised in the following table.

Table 6.1 Partial factors for checking bending moments and transverse shear forces using numerical modelling[1].

Parameter	Factor	SLS	Reliability class in ULS		
			RC1	RC2	RC3
Permanent load, g	$\gamma_{f;g}$	1.0	1.0	1.0	1.0
Variable surcharge load, p	$\gamma_{f;q}$	1.0	1.17	1.3	1.43
Tangent of internal friction angle, tan φ'	$\gamma_{m;\varphi'}$	1.0	1.2	1.25	1.3
Effective cohesion, c'	$\gamma_{m;c}$	1.0	1.3	1.45	1.6
Unit weight γ	$\gamma_{m;\gamma}$	1.0	1.0	1.0	1.0
Stiffness of embankment and subsoil	$\gamma_{m;k}$	1.0	1.0	1.0	1.0
Stiffness of geosynthetic, piles	$\gamma_{m;J}$	1.0	1.0	1.0	1.0

[1] *Based on [8], the Dutch National Appendix: Table A.3 Combination A2 Other; Table A.4a Combination M2 Overall Stability (and Table A.4c Set M1).*

7 Transition zones

7.1 The edges of the basal reinforced piled embankment (overhang)

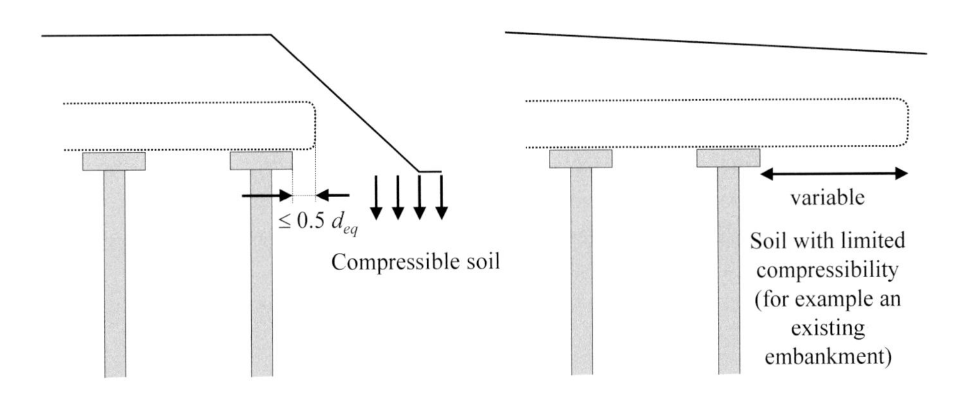

Fig.7.1 Overhang at edge of reinforced embankment.

Without additional calculations, the overhang over the edge of the pile cap of the last pile row should be at most $0.5 \cdot d_{eq}$; see Fig.7.1. For overhangs larger than $0.5 \cdot d_{eq}$, it should be demonstrated that the subgrade reaction of the surrounding soil is sufficient and that no unacceptable pile loads or deformations of the pavement structure will occur. At the transition zone from a piled embankment to a traditional embankment, the overhang may be extended for a smooth transition (see also Chapter 7.3). This is generally permissible because the embankment is subject to a residual settlement requirement and this settlement will be limited.

7.2 Connection of a piled embankment to a structure

A piled embankment and a structure on piles have different dynamic stiffnesses, which may lead to differential settlement under dynamic loading. To minimise this, a transitional approach slab may be used, comparable to the approach slabs of a viaduct.

At the transition from a piled embankment to a founded structure, the last row of piles should be situated as close as possible to the structure. The piled embankment is built up against the structure and should be terminated with a wrapped-back reinforcement layer. This solution carries the minor risk that excess aggregate material may escape from the reinforced embankment which could lead to subsidence in the road surface.

As an alternative, the last field may be left resting on the structure on an edge beam along the wall of the structure, or in combination with a foundation slab at the foot of the structure. The support area should be large enough to transfer the forces frictionally.

7.3 Transition of a piled embankment to a conventional embankment

The extent to which the transition between a piled and a conventional embankment presents a problem depends on the relative settlement that is still to occur in the conventional embankment. Techniques to ensure the transition between the two types of embankment occurs 'smoothly' are:

- position the piles gradually further apart;
- gradually drive the piles to less depth;
- allow the reinforced embankment to continue further (without piles);
- install an approach slab on the last row of piles;
- regular maintenance to compensate differential settlements;
- reduce relative settlement of the transition embankment, for example with a temporary surcharge in combination with vertical drainage or lightweight fill.

Installing the piles to less depth is to be preferred over increasing their mutual separation as the force in the geosynthetic reinforcement then does not increase. The pile layout moreover does not need to be modified which benefits consistency during construction. The pile toe depth may be reduced by 0.25 to 0.50 m per row until the piles only enter 0.50 m into the firm stratum. Settlement calculations are of course somewhat imprecise, so the risk can exist of a non-smooth transition from the piled to the conventional embankment.

Allowing the reinforced embankment to continue may be interpreted as the installation of a 'flexible' approach slab. An approach slab on the last row of piles resembles the traditional solution for viaducts and bridges. These solutions have the additional advantage that the bending moments in the piles at the edge of the piled embankment system are reduced.

8 Construction details

8.1 The construction of a piled embankment

The construction of a basal reinforced piled embankment usually proceeds as followed:

- The worksite is made accessible to the equipment to be used;
  - For example, construct a working platform of 0.20 to 0.50 m of crushed demolition waste aggregate or 0.5 to 1.0 m of sand or planks;
- Install piles or columns;
- Install pile caps or finish the pile or column caps;
- Finish flat at the top of the pile caps by backfilling the spaces between the pile caps. For this, compressible material or sand is used which is not compacted. The geosynthetic reinforcement is laid on top of this layer;
- Possibly apply material to protect the geosynthetic reinforcement from damage by the pile caps;
- If the pile caps are covered, their locations should be marked. This is necessary to be able to install the geosynthetic overlaps above the pile caps;
  - Install the geosynthetic reinforcement in accordance with the engineering drawings; in the transverse and longitudinal directions, without folds, creases or undulations, but not pretensioned;
- Apply granular fill;
  - A small amount first to keep the reinforcement in place;
  - Apply and compact the first layer of fill (for example 0.4 m of crushed demolition waste aggregate). This material should not be tipped directly on to the geosynthetic reinforcement out of the truck. The truck should tip its load on the already realised work after which the material should be spread out evenly over the geosynthetic reinforcement.
    Compaction of the first layer is not very practical as the subsoil is usually soft. Use a light static roller (e.g. max. weight approx. 3000 kg). The degree of compaction of this first layer will generally not meet the normal compaction requirements. However, due to construction traffic and the application and compaction of the succeeding layers the lowest layer will be subject to some compaction during placement of the succeeding layers;
  - Apply and compact the rest of the fill in accordance with the requirements of the engineering drawings and specifications (usually to at least 98% standard Proctor density);
  - Construction equipment should not drive directly on the geosynthetic reinfor cement. At least 0.15 m of granular fill (demonstrate with a check calculation) but

preferably 0.40 m of fill is required before construction equipment may drive on the layer;

- Once the first layer of fill is placed, the reinforcement along the edges of the reinforced embankment is usually wrapped back, so that the bottom layer is enclosed. Then the next layer of reinforcement (if any) may be installed, etc.
- Place sand fill, road base and pavement structure, as required.

8.2 Piles and pile caps

Any type of pile or soil column may be applied under a basal reinforced embankment, as long as the requirements can be met concerning the bearing capacity (Chapter 3), the bending moments and lateral shear forces (Chapter 6) and the stiffness (Table 4.2 sub 8). The characteristics of the subsoil and groundwater regime should be taken into account when deciding on the applicability of a type of bearing pile or –column. Furthermore, objects or structures in close proximity may limit the applicability of piling systems that cause vibrations or soil displacement. The quality and integrity of the bearing piles or –columns should be guaranteed to fulfil the aforementioned requirements during the design life time of the embankment.

The pile cap often consists of a square concrete slab with rounded edges on top and at the sides, and a recess in the bottom into which the pile fits. A load-distributing layer is placed between the piles and pile caps. With cast-in-situ piles, moulds are used on top of the piles that are filled with concrete along with the piles.

The piles are often end-bearing in a stratum with sufficient load bearing capacity. This should be checked by assessing the driving resistance during piling (e.g. by performing a blow count analysis), or for cast-in-situ piles or non-driven piles, by means of measuring the applied loading force.

It may occur that piles have insufficient load-bearing capacity due to variations in the soil structure. To prevent problems arising from this an extra pile may be installed near to the one with inadequate load capacity. It should then be determined whether additional geosynthetic reinforcement is required with this new, modified pile layout.

8.3 Reinforced embankment

8.3.1 Overlaps

A roll of reinforcement material is usually 5 m wide and 100 to 300 m long. Commonly, uniaxial reinforcement is used. Here, the reinforcement has its greater strength in the longitudinal direction. Fig. 8.1 and Fig. 8.2 show how overlaps should be installed.

The bottom layer of reinforcement (Fig. 8.1) lies with its strong direction perpendicular to the road axis and should always consist of a single piece across the width of the embankment.

The top layer of reinforcement (Fig. 8.2) lies with its strong direction parallel to the road axis so that overlaps are only necessary if the road is longer than the individual roll length.

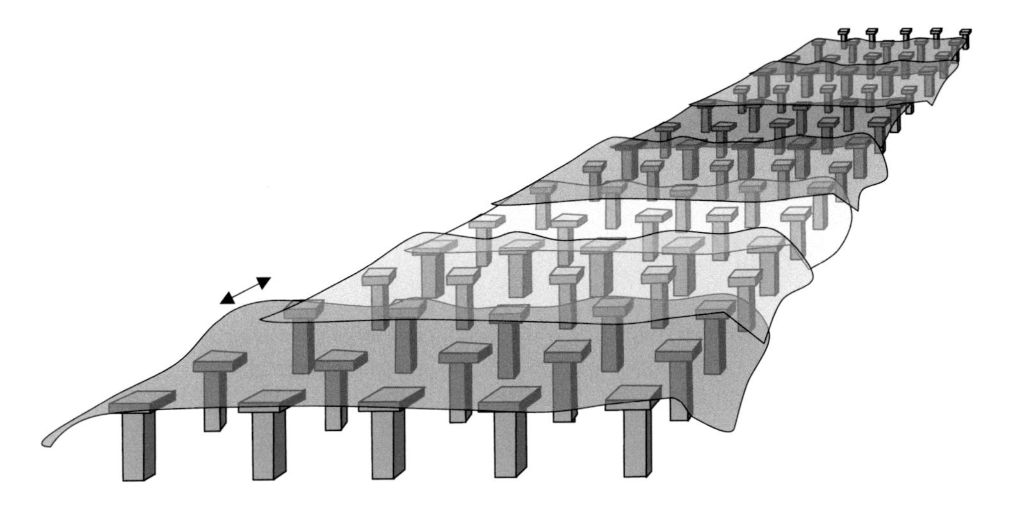

Fig. 8.1 Overlaps in transverse reinforcement.

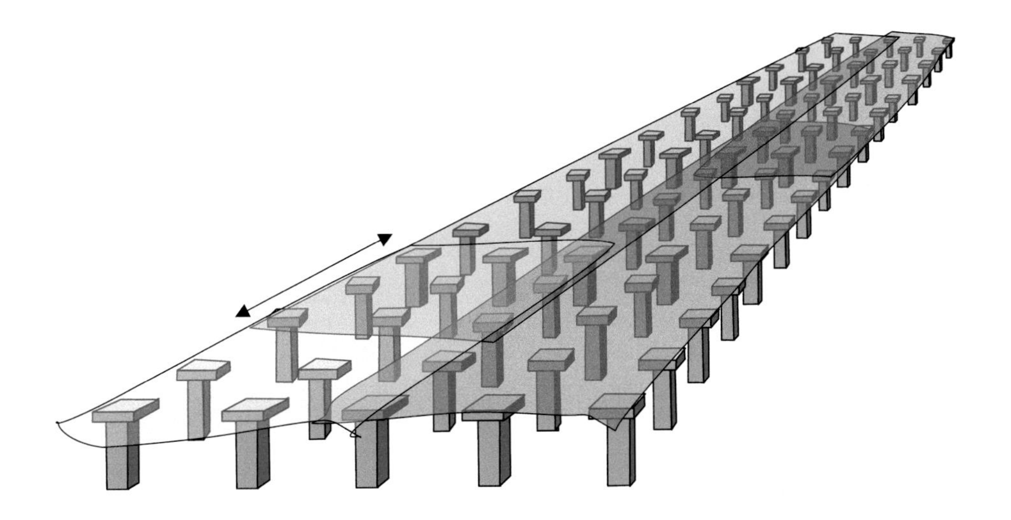

Fig. 8.2 Overlaps in longitudinal reinforcement.

An overlap in the GR's strong direction should:

- have a length that is determined by calculation;
- always be at least the length of two entire spans to cover at least three rows of piles.

An overlap in the GR's weak direction should:

- be applied above the pile caps;

- be applied over the entire width of the pile cap, unless it is demonstrated by calculation that a smaller overlap is possible. The minimum overlap should be 0.2 m.

Effects of any transverse contraction in the reinforcement material should be checked.

Overlaps can be replaced by stitched seams if feasible. The short term and long term effects of the reduction in the reinforcement's strength and the change in its strain behaviour should be included in the calculation.

8.3.2 Straight road

Extra width should be included in the roll width to enable compensation for gradual misalignment of the roll during installation. This is especially important in the longitudinal direction as the roll length is 100 - 300 m, and folds in the reinforcement layer after installation should be avoided.

Minor corrections during unrolling of the reinforcement to compensate for slight angular errors may only be introduced above the piles.

8.3.3 Bends

Fig. 8.3 shows an example of a bend with a large radius. The reinforcement layers should be cut off, and should not be laid around the corner with a pleat above the pile caps. The reinforcement layers should overlap by at least two pile spans.

The angular turns (Fig. 8.3) in the reinforcement should not be installed too close together. There should be at least one normal span between each turn, so an angle wedge may be installed in every other span. It is better to work with larger intermediate sections.

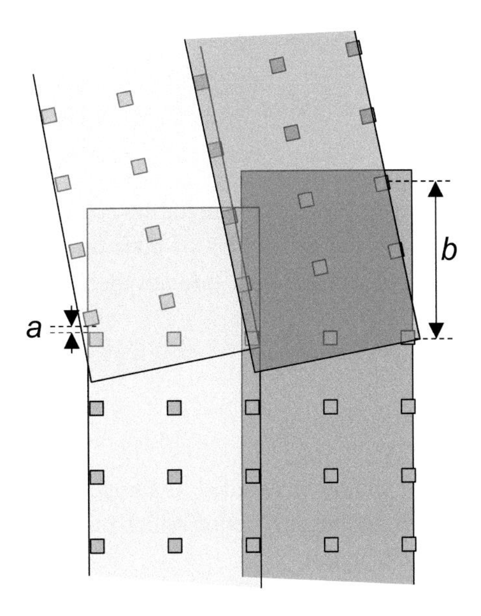

Fig.8.3 Recommended layout for bends with a large radius; *a*: at least 0.5 m, *b*: at least 2 span widths.

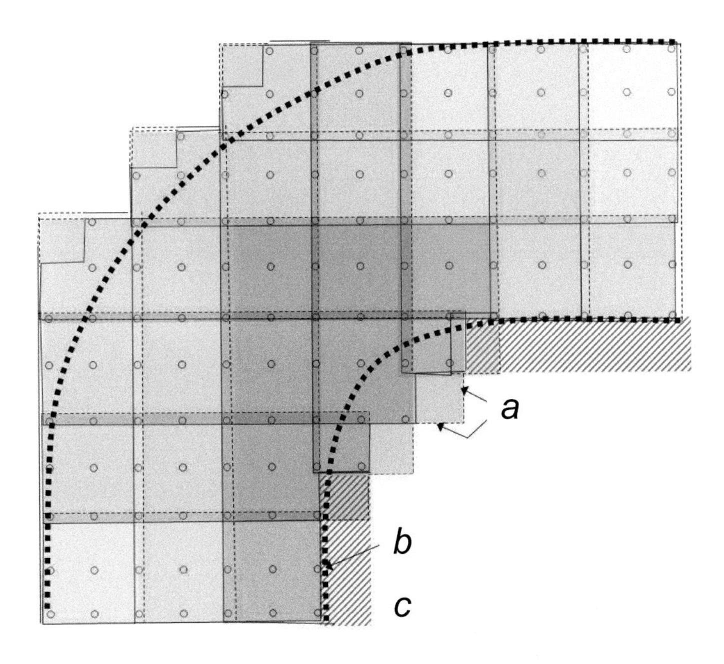

Fig.8.4 Recommended layout for bends with a small radius *a*: broken lines indicate separate pieces of reinforcement that extend outside the reinforced embankment; *b*: toe of the slope; *c* (hatched): zone with an extra row of piles for a shallower bend.

Fig. 8.4 shows an example of a sharp bend. A row of extra piles installed on the inside of the bend should always be one span longer than is theoretically necessary for the bend. This is necessary to create sufficient side enclosure. The reinforcement should be wrapped back at both the start and end of the extension (indicated with 'a' in Fig. 8.4) and in both the longitudinal and transverse directions.

At a road roundabout or intersection, a large square pile layout should be created starting from the nominal road widths and expanding with the extra rows of piles on the insides of the bends. It is better to continue the reinforced embankment through the centre of the roundabout so that no central hole is created.

8.3.4 Vertical and horizontal openings

Vertical openings through the reinforcement layer should be avoided. If an opening is made through the reinforcement the span concerned will no longer function which may lead to:

- local settlement that demands frequent maintenance;
- horizontal loads on the piles not accounted for in the design;
- eccentric loads on the edges of the reinforced embankment.

In some cases, however, vertical openings are unavoidable; Fig. 8.5 shows an example.

Fig. 8.5 Example of pile openings in reinforced embankment.

Pipes, culverts and cable paths in the reinforced embankment should also be avoided. The arching could disappear, or not recover properly, after excavation for the construction or maintenance of these elements, or due to the dynamic forces in the pipes or cables. This could have severe consequences for the reinforced embankment. For further detail, see Chapter 9.2.

Road signs etc., which cause limited disruption, may be installed in holes drilled through the reinforced embankment. A small, stable hole is permissible, but no large holes should be created. It is always better to avoid disruptions to the reinforced embankment. Large, heavy road signs should be installed outside the alignment of the reinforced embankment with their own individual foundations.

8.3.5 Enclosure of the reinforced embankment at the sides

The mattress is often constructed using a temporary formwork at the sides. After installing the formwork (see Fig. 8.6), the two reinforcement layers are installed inside and then the fill is placed and contained within the side enclosure. Following this, the reinforcement material is folded back and another layer of fill is placed (as a cover layer at least 0.15 m thick). Once completed, the formwork is then removed.

It is also possible that no side enclosure of the reinforced embankment (folding back of the reinforcement) is used. In this case it should then be demonstrated that the force transfer at the toe of the slope from the reinforcement layer to the piles and from the fill to the piles is sufficient.

Fig. 8.6 Piled embankment of the N210 at the Krimpenerwaard, The Netherlands; formwork to temporarily enclose the aggregate before the geosynthetic is wrapped back.

Fig. 8.7 Layer-by-layer filling of a piled embankment at Hafencity, Hamburg, Germany.

9 Management and maintenance

9.1 General

Due to the nearly-settlement-free nature of a piled embankment, the quality of the pavement structure on top of the embankment is enhanced, thus requiring less maintenance.

Maintenance to the piles and reinforced mattress is difficult and there should be restrictions on future works or structures to be located in the vicinity of the piled embankment. These might be future embankment widening or the installation of cables and pipelines. It is therefore crucial to formulate a maintenance strategy for the structure's entire service life during the design and planning phase. This last point falls outside the scope of this guideline. This chapter considers a number of management and maintenance details.

9.2 Cables and pipelines

For transverse connections already present, or to be installed during the construction of the piled embankment, custom work is usually required, for example, by means of overcapping. Less vulnerable cables and pipelines may be installed between the piles by designing the pile arrangement to accommodate them or by diverting the connection.

Subsequent installation of connections beneath an excisting piled embankment is difficult unless measures have been taken in advance. Here, a distinction should be made between:

- Major infrastructure, such as main sewers, pressurised pipelines, etc. Facilities can be incorporated in advance to facilitate this, such as the inclusion at regular intervals of empty ducts or cable tunnels. Alternatively, horizontally-directed drilling could be conducted beneath the piles. It is already common practice to drill beneath vertical drains, which is comparable to this situation.
- Cables and pipelines that are usually installed in the uppermost metre of sand. It is relatively straightforward to account for these by applying a finishing layer of at least 1 m of sand.

Pipes and sewers for rainwater drainage and other elements parallel to the road axis should be installed outside the reinforced embankment or in the toe of the slope.

9.3 Embankment widening

In general, raising or lowering of the surface level immediately adjacent to a piled embankment should not be allowed because this could lead to horizontal soil deformations that might exert an unsafe load on the piles. As good an account as possible should be taken of these during the design of the piled embankment. Any future structures should be designed such that they cannot exert a detrimental effect on the embankment piles. It should be assessed for each case what is permissible.

9.4 Service life and function retention of materials to be used

Inspection of the materials used may be desirable during the service life. One option is to include non-functional pieces of geosynthetic reinforcement that are subjected to the same conditions as the 'functional' geosynthetic reinforcements. By retrieving, inspecting and testing these sample pieces of geosynthetic, information may be obtained about the state of the rest of the geosynthetic reinforcements present in the embankment. In this way, the possibility of extending the service life at the end of the design service life can be evaluated.

9.5 Other management aspects

9.5.1 Coverage of geosynthetic reinforcement in connection with maintenance of slopes

During maintenance on the verges and slopes of rail and road beds, account should be taken of the presence of the geosynthetic reinforcement. The reinforcement should have been installed at a depth giving a certain minimum coverage so that it cannot be affected or damaged by maintenance work.

9.5.2 Accidents

Hazardous substances may be released during accidents that could affect the geosynthetics used. It is important in case of accidents to investigate whether the geosynthetics have been affected. Consequently, it may be assessed to what extent the geosynthetics remain functional and still fulfil the requirements imposed. Non functional pieces of geosynthetic described in Chapter 9.4 could be used for this inspection.

9.5.3 Damage to geosynthetic reinforcements

Chapter 2.10.1 considers the reduction factors that take into account damage to the geosynthetic reinforcement by the environment in which it is placed. There should be a check whether the conditions in the field correspond to the assumed design conditions. Moreover, the road manager should check for damage by rodents, termites, etc.

9.5.4 Exceeding the deformation requirements

If the deformation requirements are exceeded and/or inadequate function of the basal reinforced piled embankment is observed repair works should be conducted. Care should be taken to ensure that any repair works do not impair the structure further or even demolish it partially.

Appendix A
Traffic load tables

Table A.1 Maximum average uniform traffic load $p_{traffic}$ on piled embankment based on NEN-EN 1991-2 | N = 200,000 for lane 1, heavy traffic (F_{wheel} = 116 kN and $q_{uniform}$ = 7.0 kPa).

H_{eq} (m)	$p_{traffic;max}$ (kPa)	$p_{traffic}$ (kPa) for pile arrangement (m^2)					
		0.5 x 0.5	1.0 x 1.0	1.5 x 1.5	2.0 x 2.0	2.5 x 2.5	3.0 x 3.0
1.00	66.20	61.75	56.03	51.71	45.92	44.59	41.36
1.20	53.45	51.30	48.86	45.72	41.80	40.34	37.68
1.40	45.55	44.65	43.04	40.89	38.17	36.67	34.42
1.60	40.20	39.60	38.42	36.85	34.91	33.45	31.51
1.80	35.88	35.40	34.54	33.59	31.96	30.60	28.92
2.00	32.29	32.00	31.44	30.89	29.29	28.05	26.59
2.20	29.42	29.23	28.90	28.44	26.87	25.77	24.50
2.40	27.06	26.91	26.61	26.15	24.69	23.71	22.61
2.60	24.91	24.77	24.48	24.04	22.72	21.87	20.91
2.80	22.92	22.79	22.52	22.12	20.95	20.20	19.38
3.00	21.10	20.98	20.74	20.38	19.35	18.70	17.99
3.20	19.45	19.34	19.13	18.82	17.92	17.35	16.73
3.40	17.96	17.86	17.68	17.40	16.62	16.13	15.58
3.60	16.62	16.53	16.37	16.13	15.45	15.02	14.55
3.80	15.41	15.34	15.19	14.99	14.39	14.02	13.61
4.00	14.32	14.26	14.13	13.95	13.43	13.11	12.75
4.20	13.34	13.28	13.18	13.02	12.57	12.28	11.97
4.40	12.45	12.41	12.31	12.17	11.78	11.53	11.25
4.60	11.65	11.61	11.53	11.41	11.06	10.84	10.60
4.80	10.93	10.89	10.82	10.71	10.41	10.22	10.00
5.00	10.27	10.24	10.17	10.08	9.81	9.64	9.45
5.20	9.67	9.64	9.58	9.50	9.26	9.11	8.94
5.40	9.12	9.09	9.04	8.97	8.76	8.63	8.48
5.60	8.62	8.60	8.55	8.49	8.30	8.18	8.05

H_{eq} (m)	$p_{traffic;max}$ (kPa)	$p_{traffic}$ (kPa) for pile arrangement (m^2)					
		0.5 x 0.5	1.0 x 1.0	1.5 x 1.5	2.0 x 2.0	2.5 x 2.5	3.0 x 3.0
5.80	8.16	8.14	8.10	8.04	7.88	7.77	7.65
6.00	7.74	7.72	7.68	7.63	7.49	7.39	7.28
6.20	7.35	7.33	7.30	7.26	7.12	7.04	6.94
6.40	6.99	6.98	6.95	6.91	6.79	6.71	6.63
6.60	6.66	6.65	6.62	6.58	6.48	6.41	6.33
6.80	6.35	6.34	6.32	6.29	6.19	6.13	6.06
7.00	6.07	6.06	6.04	6.01	5.92	5.86	5.80
7.20	5.80	5.79	5.78	5.75	5.67	5.62	5.56
7.40	5.56	5.55	5.53	5.51	5.44	5.39	5.34
7.60	5.33	5.32	5.30	5.28	5.22	5.18	5.13
7.80	5.11	5.10	5.09	5.07	5.01	4.97	4.93
8.00	4.91	4.90	4.89	4.87	4.82	4.79	4.75

$p_{traffic,max}$ is the maximum vertical load on the GR;

$p_{traffic}$ is the maximum average load on the GR. This value is used as the characteristic value of the traffic load p and to determine the value of κ for the reduction of the arching. Possibly, a permanent load may be added to this, so that the design load p becomes: $p = p_{traffic} + p_{permanent}$.

Table A.2 Maximum average uniform traffic load $p_{traffic}$ on piled embankment based on NEN-EN 1991-2 | N = 20,000 for lane 1, heavy traffic (F_{wheel} = 114 kN and $q_{uniform}$ = 6.8 kPa).

H_{eq} (m)	$p_{traffic;max}$ (kPa)	$p_{traffic}$ (kPa) for pile arrangement (m^2)					
		0.5 x 0.5	1.0 x 1.0	1.5 x 1.5	2.0 x 2.0	2.5 x 2.5	3.0 x 3.0
1.00	65.04	60.67	55.05	50.79	45.10	43.80	40.63
1.20	52.51	50.39	47.99	44.91	41.06	39.62	37.01
1.40	44.74	43.86	42.28	40.16	37.49	36.02	33.81
1.60	39.49	38.90	37.74	36.19	34.28	32.86	30.95
1.80	35.25	34.78	33.93	32.99	31.39	30.06	28.40
2.00	31.72	31.43	30.88	30.34	28.76	27.55	26.12
2.20	28.90	28.71	28.39	27.93	26.39	25.31	24.06
2.40	26.57	26.43	26.14	25.68	24.25	23.29	22.21
2.60	24.47	24.33	24.04	23.61	22.32	21.48	20.54
2.80	22.51	22.38	22.12	21.73	20.58	19.84	19.03
3.00	20.72	20.60	20.37	20.02	19.01	18.37	17.66

H_{eq} (m)	$p_{traffic;max}$ (kPa)	$p_{traffic}$ (kPa) for pile arrangement (m^2)					
		0.5 x 0.5	1.0 x 1.0	1.5 x 1.5	2.0 x 2.0	2.5 x 2.5	3.0 x 3.0
3.20	19.10	19.00	18.79	18.48	17.59	17.04	16.43
3.40	17.64	17.54	17.36	17.09	16.32	15.84	15.30
3.60	16.32	16.24	16.08	15.84	15.17	14.75	14.29
3.80	15.13	15.06	14.92	14.72	14.13	13.77	13.36
4.00	14.06	14.00	13.88	13.70	13.19	12.87	12.52
4.20	13.10	13.05	12.94	12.78	12.34	12.06	11.75
4.40	12.23	12.18	12.09	11.95	11.57	11.32	11.05
4.60	11.44	11.40	11.32	11.20	10.86	10.65	10.41
4.80	10.73	10.69	10.62	10.52	10.22	10.03	9.82
5.00	10.08	10.05	9.99	9.90	9.63	9.47	9.28
5.20	9.49	9.46	9.41	9.33	9.10	8.95	8.78
5.40	8.95	8.93	8.88	8.81	8.60	8.47	8.32
5.60	8.46	8.44	8.40	8.33	8.15	8.03	7.90
5.80	8.01	7.99	7.95	7.90	7.73	7.63	7.51
6.00	7.60	7.58	7.54	7.49	7.35	7.26	7.15
6.20	7.21	7.20	7.17	7.12	6.99	6.91	6.82
6.40	6.86	6.85	6.82	6.78	6.66	6.59	6.51
6.60	6.54	6.52	6.50	6.46	6.36	6.29	6.22
6.80	6.24	6.22	6.20	6.17	6.08	6.02	5.95
7.00	5.96	5.95	5.93	5.90	5.81	5.76	5.69
7.20	5.70	5.69	5.67	5.64	5.57	5.52	5.46
7.40	5.45	5.45	5.43	5.41	5.34	5.29	5.24
7.60	5.23	5.22	5.21	5.19	5.12	5.08	5.03
7.80	5.02	5.01	5.00	4.98	4.92	4.88	4.84
8.00	4.82	4.81	4.80	4.78	4.73	4.70	4.66

$p_{traffic,max}$ is the maximum vertical load on the GR;

$p_{traffic}$ is the maximum average load on the GR. This value is used as the characteristic value of the traffic load p and to determine the value of κ for the reduction of the arching. Possibly, a permanent load may be added to this, so that the design load p becomes: $p = p_{traffic} + p_{permanent}$.

Table A.3 Maximum average uniform traffic load $p_{traffic}$ on piled embankment based on NEN-EN 1991-2 | N = 2,000 for lane 1, heavy traffic (F_{wheel} = 109 kN and $q_{uniform}$ = 6.6 kPa).

H_{eq} (m)	$p_{traffic;max}$ (kPa)	$p_{traffic}$ (kPa) for pile arrangement (m²)					
		0.5 x 0.5	1.0 x 1.0	1.5 x 1.5	2.0 x 2.0	2.5 x 2.5	3.0 x 3.0
1.00	62.20	58.02	52.64	48.58	43.14	41.89	38.86
1.20	50.22	48.19	45.90	42.95	39.27	37.89	35.40
1.40	42.79	41.95	40.43	38.41	35.86	34.45	32.33
1.60	37.77	37.20	36.09	34.62	32.79	31.43	29.60
1.80	33.71	33.26	32.45	31.55	30.02	28.75	27.17
2.00	30.34	30.06	29.54	29.02	27.51	26.35	24.98
2.20	27.64	27.46	27.15	26.72	25.24	24.21	23.01
2.40	25.42	25.28	25.00	24.56	23.19	22.28	21.24
2.60	23.40	23.27	23.00	22.58	21.35	20.54	19.64
2.80	21.53	21.40	21.16	20.78	19.68	18.98	18.20
3.00	19.82	19.71	19.48	19.15	18.18	17.57	16.89
3.20	18.27	18.17	17.97	17.68	16.83	16.30	15.71
3.40	16.87	16.78	16.61	16.35	15.61	15.15	14.64
3.60	15.61	15.53	15.38	15.15	14.51	14.11	13.67
3.80	14.47	14.41	14.27	14.08	13.52	13.17	12.78
4.00	13.45	13.39	13.28	13.11	12.62	12.31	11.98
4.20	12.53	12.48	12.38	12.23	11.80	11.54	11.24
4.40	11.70	11.65	11.56	11.44	11.06	10.83	10.57
4.60	10.95	10.91	10.83	10.72	10.39	10.18	9.96
4.80	10.26	10.23	10.16	10.06	9.78	9.59	9.39
5.00	9.64	9.61	9.55	9.47	9.21	9.05	8.88
5.20	9.08	9.05	9.00	8.92	8.70	8.56	8.40
5.40	8.56	8.54	8.50	8.43	8.23	8.10	7.96
5.60	8.09	8.07	8.03	7.97	7.80	7.69	7.56
5.80	7.66	7.64	7.61	7.55	7.40	7.30	7.19
6.00	7.27	7.25	7.22	7.17	7.03	6.94	6.84
6.20	6.90	6.89	6.86	6.82	6.69	6.61	6.52
6.40	6.52	6.55	6.53	6.49	6.38	6.30	6.22
6.60	6.25	6.24	6.22	6.18	6.08	6.02	5.95
6.80	5.97	5.95	5.93	5.90	5.81	5.75	5.69

H_{eq} (m)	$p_{traffic;max}$ (kPa)	$p_{traffic}$ (kPa) for pile arrangement (m²)					
		0.5 x 0.5	1.0 x 1.0	1.5 x 1.5	2.0 x 2.0	2.5 x 2.5	3.0 x 3.0
7.00	5.70	5.69	5.67	5.64	5.56	5.51	5.45
7.20	5.45	5.44	5.42	5.40	5.32	5.28	5.22
7.40	5.22	5.21	5.19	5.17	5.11	5.06	5.01
7.60	5.00	5.00	4.98	4.96	4.90	4.86	4.82
7.80	4.80	4.79	4.78	4.76	4.71	4.67	4.63
8.00	4.61	4.61	4.59	4.58	4.53	4.49	4.46

$p_{traffic,max}$ is the maximum vertical load on the GR;

$p_{traffic}$ is the maximum average load on the GR. This value is used as the characteristic value of the traffic load p and to determine the value of κ for the reduction of the arching. Possibly, a permanent load may be added to this, so that the design load p becomes: $p = p_{traffic} + p_{permanent}$.

Table A.4 Maximum average uniform traffic load $p_{traffic}$ on piled embankment based on NEN-EN 1991-2 | $N = 200$ for lane 1, heavy traffic ($F_{wheel} = 106$ kN and $q_{uniform} = 6.3$ kPa).

H_{eq} (m)	$p_{traffic;max}$ (kPa)	$p_{traffic}$ (kPa) for pile arrangement (m²)					
		0.5 x 0.5	1.0 x 1.0	1.5 x 1.5	2.0 x 2.0	2.5 x 2.5	3.0 x 3.0
1.00	60.45	56.39	51.16	47.21	41.92	40.71	37.75
1.20	48.80	46.84	44.60	41.74	38.16	36.82	34.40
1.40	41.58	40.76	39.29	37.33	34.84	33.47	31.41
1.60	36.70	36.15	35.07	33.64	31.86	30.53	28.76
1.80	32.75	32.32	31.53	30.66	29.17	27.93	26.39
2.00	29.48	29.21	28.70	28.20	26.73	25.60	24.27
2.20	26.85	26.68	26.38	25.96	24.52	23.52	22.36
2.40	24.69	24.56	24.29	23.87	22.53	21.64	20.63
2.60	22.73	22.60	22.34	21.94	20.74	19.95	19.08
2.80	20.92	20.80	20.55	20.19	19.12	18.44	17.68
3.00	19.25	19.14	18.93	18.60	17.66	17.07	16.41
3.20	17.75	17.65	17.46	17.17	16.35	15.83	15.26
3.40	16.39	16.30	16.13	15.88	15.16	14.71	14.22
3.60	15.16	15.09	14.94	14.72	14.09	13.70	13.27
3.80	14.06	13.99	13.86	13.67	13.13	12.79	12.41
4.00	13.06	13.01	12.90	12.73	12.26	11.96	11.63
4.20	12.17	12.12	12.02	11.88	11.46	11.20	10.92

H_{eq} (m)	$p_{traffic;max}$ (kPa)	$p_{traffic}$ (kPa) for pile arrangement (m^2)					
		0.5 x 0.5	1.0 x 1.0	1.5 x 1.5	2.0 x 2.0	2.5 x 2.5	3.0 x 3.0
4.40	11.36	11.32	11.23	11.11	10.74	10.52	10.26
4.60	10.63	10.59	10.52	10.41	10.09	9.89	9.67
4.80	9.97	9.93	9.87	9.77	9.49	9.32	9.12
5.00	9.37	9.34	9.28	9.19	8.95	8.79	8.62
5.20	8.82	8.79	8.74	8.67	8.45	8.31	8.16
5.40	8.32	8.29	8.25	8.18	7.99	7.87	7.73
5.60	7.86	7.84	7.80	7.74	7.57	7.46	7.34
5.80	7.44	7.42	7.39	7.34	7.18	7.09	6.98
6.00	7.06	7.04	7.01	6.96	6.83	6.74	6.64
6.20	6.70	6.69	6.66	6.62	6.50	6.42	6.33
6.40	6.37	6.36	6.34	6.30	6.19	6.12	6.04
6.60	6.07	6.06	6.04	6.00	5.91	5.84	5.77
6.80	5.79	5.78	5.76	5.73	5.64	5.59	5.52
7.00	5.53	5.52	5.50	5.48	5.40	5.35	5.29
7.20	5.29	5.28	5.27	5.24	5.17	5.12	5.07
7.40	5.07	5.06	5.04	5.02	4.96	4.91	4.87
7.60	4.86	4.85	4.84	4.82	4.76	4.72	4.67
7.80	4.66	4.65	4.64	4.62	4.57	4.53	4.49
8.00	4.48	4.47	4.46	4.44	4.39	4.36	4.33

$p_{traffic,max}$ is the maximum vertical load on the GR;

$p_{traffic}$ is the maximum average load on the GR. This value is used as the characteristic value of the traffic load p and to determine the value of κ for the reduction of the arching. Possibly, a permanent load may be added to this, so that the design load p becomes: $p = p_{traffic} + p_{permanent}$.

Table A.5 Maximum average uniform traffic load $p_{traffic}$ on piled embankment based on NEN-EN 1991-2 | N = 200,000 for lane 1, heavy traffic (F_{wheel} = 116 kN and quniform = 7.0 kPa) and lane 2 (F_{wheel} = 95 kN and $q_{uniform}$ = 2.4 kPa).

H_{eq} (m)	$p_{traffic;max}$ (kPa)	$p_{traffic}$ (kPa) for pile arrangement (m²)					
		0.5 x 0.5	1.0 x 1.0	1.5 x 1.5	2.0 x 2.0	2.5 x 2.5	3.0 x 3.0
1.00	78.27	73.91	72.48	68.29	60.01	50.98	45.67
1.20	67.82	66.55	64.74	60.55	54.20	47.48	42.71
1.40	61.06	60.05	57.94	54.21	49.27	44.16	40.06
1.60	55.04	54.07	52.03	48.90	45.01	41.06	37.52
1.80	49.59	48.70	46.93	44.37	41.34	38.17	35.20
2.00	44.76	44.01	42.54	40.50	38.08	35.50	32.98
2.20	40.57	39.96	38.77	37.12	35.16	33.04	30.90
2.40	36.95	36.45	35.49	34.15	32.54	30.77	28.96
2.60	33.80	33.39	32.60	31.50	30.17	28.69	27.15
2.80	31.05	30.71	30.06	29.15	28.04	26.79	25.46
3.00	28.62	28.34	27.81	27.04	26.11	25.04	23.89
3.20	23.89	26.48	26.24	25.79	24.35	23.44	22.44
3.40	24.57	24.37	23.99	23.44	22.76	21.97	21.10
3.60	22.86	22.69	22.36	21.89	21.30	20.62	19.86
3.80	21.32	21.18	20.89	20.49	19.98	19.38	18.71
4.00	19.93	19.81	19.56	19.21	18.76	18.24	17.65
4.20	18.67	18.56	18.35	18.04	17.65	17.19	16.67
4.40	17.53	17.43	17.25	16.98	16.63	16.22	15.76
4.60	16.48	16.40	16.24	16.00	15.69	15.33	14.92
4.80	15.53	15.46	15.31	15.10	14.83	14.51	14.14
5.00	14.66	14.59	14.46	14.28	14.03	13.74	13.41
5.20	13.85	13.79	13.68	13.51	13.30	13.04	12.74
5.40	13.11	13.06	12.96	12.81	12.62	12.38	12.12
5.60	12.43	12.38	12.29	12.16	11.99	11.78	11.53
5.80	11.80	11.76	11.68	11.56	11.40	11.21	10.99
6.00	11.22	11.18	11.10	11.00	10.85	10.68	10.49
6.20	10.67	10.64	10.57	10.48	10.35	10.19	10.01
6.40	10.17	10.14	10.08	9.99	9.88	9.74	9.57
6.60	9.70	9.68	9.62	9.54	9.44	9.31	9.16
6.80	9.27	9.24	9.19	9.12	9.02	8.91	8.77

H_{eq} (m)	$p_{traffic;max}$ (kPa)	$p_{traffic}$ (kPa) for pile arrangement (m^2)					
		0.5 x 0.5	1.0 x 1.0	1.5 x 1.5	2.0 x 2.0	2.5 x 2.5	3.0 x 3.0
7.00	8.86	8.84	8.79	8.73	8.64	8.53	8.41
7.20	8.48	8.46	8.42	8.36	8.28	8.18	8.07
7.40	8.12	8.10	8.07	8.01	7.94	7.85	7.75
7.60	7.79	7.77	7.74	7.69	7.62	7.54	7.44
7.80	7.48	7.46	7.43	7.38	7.32	7.25	7.16
8.00	7.18	7.17	7.14	7.10	7.04	6.97	6.89

$p_{traffic,max}$ is the maximum vertical load on the GR;

$p_{traffic}$ is the maximum average load on the GR. This value is used as the characteristic value of the traffic load p and to determine the value of κ for the reduction of the arching. Possibly, a permanent load may be added to this, so that the design load p becomes: $p = p_{traffic} + p_{permanent}$

Table A.6 Maximum average uniform traffic load $p_{traffic}$ on piled embankment based on NEN-EN 1991-2 | N = 20,000 for lane 1, heavy traffic (F_{wheel} = 114 kN and $q_{uniform}$ = 6.8 kPa) and lane 2 (F_{wheel} = 88 kN and $q_{uniform}$ = 2.2 kPa).

H_{eq} (m)	$p_{traffic;max}$ (kPa)	$p_{traffic}$ (kPa) for pile arrangement (m²)					
		0.5 x 0.5	1.0 x 1.0	1.5 x 1.5	2.0 x 2.0	2.5 x 2.5	3.0 x 3.0
1.00	76.50	72.22	70.78	66.67	58.54	49.68	44.50
1.20	66.25	65.00	63.19	59.08	52.85	46.26	41.60
1.40	59.62	58.62	56.54	52.88	48.03	43.02	39.01
1.60	53.72	52.75	50.76	47.68	43.88	39.99	36.53
1.80	48.37	47.50	45.77	43.26	40.29	37.18	34.26
2.00	43.65	42.91	41.48	39.48	37.10	34.57	32.10
2.20	39.57	38.96	37.79	36.18	34.25	32.17	30.07
2.40	36.02	35.53	34.58	33.27	31.69	29.96	28.18
2.60	32.94	32.54	31.77	30.69	29.38	27.93	26.41
2.80	30.25	29.92	29.28	28.39	27.30	26.07	24.77
3.00	27.88	27.61	27.08	26.33	25.41	24.36	23.24
3.20	25.79	25.56	25.11	24.48	23.70	22.80	21.82
3.40	23.92	23.73	23.35	22.82	22.14	21.37	20.52
3.60	22.25	22.09	21.77	21.31	20.73	20.05	19.30
3.80	20.75	20.61	20.33	19.94	19.43	18.84	18.19
4.00	19.39	19.27	19.03	18.69	18.25	17.73	17.15
4.20	18.17	18.06	17.85	17.55	17.16	16.71	16.20
4.40	17.05	16.96	16.77	16.51	16.17	15.77	15.31
4.60	16.03	15.95	15.79	15.55	15.25	14.90	14.49
4.80	15.10	15.03	14.89	14.68	14.41	14.09	13.73
5.00	14.25	14.18	14.06	13.87	13.64	13.35	13.03
5.20	13.46	13.41	13.29	13.13	12.92	12.66	12.37
5.40	12.74	12.69	12.59	12.44	12.25	12.03	11.76
5.60	12.07	12.03	11.94	11.81	11.64	11.43	11.20
5.80	11.46	11.42	11.34	11.22	11.07	10.88	10.67
6.00	10.89	10.85	10.78	10.68	10.54	10.37	10.18
6.20	10.36	10.33	10.26	10.17	10.04	9.89	9.72
6.40	9.87	9.84	9.78	9.70	9.58	9.44	9.29
6.60	9.42	9.39	9.34	9.26	9.15	9.03	8.88

H_{eq} (m)	$p_{traffic;max}$ (kPa)	$p_{traffic}$ (kPa) for pile arrangement (m^2)					
		0.5 x 0.5	1.0 x 1.0	1.5 x 1.5	2.0 x 2.0	2.5 x 2.5	3.0 x 3.0
6.80	8.99	8.97	8.92	8.85	8.75	8.64	8.51
7.00	8.59	8.57	8.53	8.46	8.38	8.27	8.15
7.20	8.22	8.20	8.16	8.10	8.03	7.93	7.82
7.40	7.88	7.86	7.82	7.77	7.70	7.61	7.51
7.60	7.55	7.54	7.50	7.45	7.39	7.31	7.21
7.80	7.25	7.23	7.20	7.16	7.10	7.02	6.94
8.00	6.96	6.95	6.92	6.88	6.82	6.75	6.67

$p_{traffic,max}$ is the maximum vertical load on the GR;

$p_{traffic}$ is the maximum average load on the GR. This value is used as the characteristic value of the traffic load p and to determine the value of κ for the reduction of the arching. Possibly, a permanent load may be added to this, so that the design load p becomes: $p = p_{traffic} + p_{permanent}$.

Table A.7 Maximum average uniform traffic load $p_{traffic}$ on piled embankment based on NEN-EN 1991-2 | $N = 2{,}000$ for lane 1, heavy traffic ($F_{wheel} = 109$ kN and $q_{uniform} = 6.6$ kPa) and lane 2 ($F_{wheel} = 81$ kN and $q_{uniform} = 2.0$ kPa).

H_{eq} (m)	$p_{traffic;max}$ (kPa)	$p_{traffic}$ (kPa) for pile arrangement (m^2)					
		0.5 x 0.5	1.0 x 1.0	1.5 x 1.5	2.0 x 2.0	2.5 x 2.5	3.0 x 3.0
1.00	72.97	68.88	67.52	63.58	55.80	47.33	42.36
1.20	63.17	61.98	60.27	56.33	50.38	44.06	39.59
1.40	56.85	55.89	53.91	50.41	45.77	40.97	37.13
1.60	51.22	50.30	48.39	45.45	41.80	38.09	34.76
1.80	46.11	45.28	43.62	41.22	38.38	35.40	32.61
2.00	41.61	40.90	39.52	37.61	35.34	32.92	30.55
2.20	37.70	37.12	36.01	34.46	32.62	30.63	28.62
2.40	34.32	33.85	32.94	31.68	30.17	28.52	26.82
2.60	31.38	31.00	30.26	29.22	27.97	26.59	25.13
2.80	28.81	28.50	27.89	27.03	25.99	24.81	23.57
3.00	26.55	26.29	25.79	25.07	24.19	23.19	22.11
3.20	24.55	24.33	23.91	23.31	22.56	21.70	20.76
3.40	22.77	22.59	22.23	21.72	21.07	20.33	19.52
3.60	21.18	21.02	20.72	20.28	19.72	19.08	18.36
3.80	19.75	19.61	19.35	18.97	18.49	17.92	17.30
4.00	18.46	18.34	18.11	17.78	17.36	16.87	16.31

H_{eq} (m)	$p_{traffic;max}$ (kPa)	$p_{traffic}$ (kPa) for pile arrangement (m^2)					
		0.5 x 0.5	1.0 x 1.0	1.5 x 1.5	2.0 x 2.0	2.5 x 2.5	3.0 x 3.0
4.20	17.29	17.18	16.98	16.70	16.33	15.89	15.40
4.40	16.22	16.13	15.96	15.70	15.38	14.99	14.56
4.60	15.25	15.17	15.02	14.79	14.51	14.17	13.78
4.80	14.36	14.29	14.16	13.96	13.70	13.40	13.05
5.00	13.55	13.49	13.37	13.19	12.96	12.69	12.38
5.20	12.80	12.75	12.64	12.48	12.28	12.04	11.76
5.40	12.11	12.07	11.97	11.83	11.65	11.43	11.18
5.60	11.48	11.44	11.35	11.22	11.06	10.86	10.64
5.80	10.89	10.85	10.78	10.66	10.52	10.34	10.14
6.00	10.35	10.32	10.25	10.14	10.01	9.85	9.67
6.20	9.85	9.82	9.75	9.66	9.54	9.40	9.23
6.40	9.38	9.35	9.30	9.21	9.10	8.97	8.82
6.60	8.95	8.92	8.87	8.79	8.69	8.57	8.44
6.80	8.54	8.52	8.47	8.40	8.31	8.20	8.08
7.00	8.16	8.14	8.10	8.04	7.95	7.86	7.74
7.20	7.81	7.79	7.75	7.69	7.62	7.53	7.42
7.40	7.48	7.46	7.43	7.37	7.31	7.22	7.13
7.60	7.17	7.15	7.12	7.07	7.01	6.93	6.85
7.80	6.88	6.87	6.84	6.79	6.73	6.66	6.58
8.00	6.61	6.59	6.57	6.53	6.47	6.41	6.33

$p_{traffic,max}$ is the maximum vertical load on the GR;

$p_{traffic}$ is the maximum average load on the GR. This value is used as the characteristic value of the traffic load p and to determine the value of κ for the reduction of the arching. Possibly, a permanent load may be added to this, so that the design load p becomes: $p = p_{traffic} + p_{permanent}$.

Table A.8 Maximum average uniform traffic load $p_{traffic}$ on piled embankment based on NEN-EN 1991-2 | N = 200 for lane 1, heavy traffic (F_{wheel} = 106 kN and $q_{uniform}$ = 6.3 kPa) and lane 2 (F_{wheel} = 74 kN and $q_{uniform}$ = 1.9 kPa).

H_{eq} (m)	$p_{traffic;max}$ (kPa)	$p_{traffic}$ (kPa) for pile arrangement (m^2)					
		0.5 x 0.5	1.0 x 1.0	1.5 x 1.5	2.0 x 2.0	2.5 x 2.5	3.0 x 3.0
1.00	70.55	66.56	65.19	61.37	53.83	45.61	40.82
1.20	61.03	59.87	58.17	54.35	48.57	42.46	38.15
1.40	54.90	53.97	52.01	48.62	44.12	39.48	35.76
1.60	49.44	48.53	46.67	43.82	40.29	36.69	33.48
1.80	44.49	43.68	42.07	39.74	36.99	34.10	31.39
2.00	40.13	39.45	38.11	36.26	34.05	31.70	29.41
2.20	36.36	35.80	34.71	33.21	31.42	29.49	27.54
2.40	33.09	32.63	31.75	30.53	29.06	27.46	25.81
2.60	30.25	29.87	29.16	28.15	26.94	25.59	24.18
2.80	27.76	27.46	26.87	26.04	25.02	23.88	22.67
3.00	25.58	25.33	24.84	24.14	23.29	22.31	21.27
3.20	23.65	23.44	23.03	22.44	21.71	20.88	19.97
3.40	21.93	21.75	21.40	20.90	20.28	19.56	18.77
3.60	20.39	20.24	19.94	19.51	18.98	18.35	17.66
3.80	19.01	18.88	18.62	18.25	17.79	17.24	16.63
4.00	17.76	17.65	17.43	17.11	16.70	16.22	15.68
4.20	16.63	16.53	16.34	16.06	15.70	15.28	14.80
4.40	15.60	15.52	15.35	15.10	14.79	14.41	13.99
4.60	14.67	14.59	14.44	14.22	13.95	13.61	13.24
4.80	13.81	13.74	13.61	13.42	13.17	12.88	12.54
5.00	13.02	12.97	12.85	12.68	12.46	12.19	11.89
5.20	12.30	12.25	12.15	12.00	11.80	11.56	11.29
5.40	11.64	11.59	11.50	11.37	11.19	10.98	10.73
5.60	11.03	10.99	10.90	10.78	10.62	10.43	10.21
5.80	10.46	10.43	10.35	10.24	10.10	9.93	9.73
6.00	9.94	9.91	9.84	9.74	9.61	9.46	9.28
6.20	9.46	9.42	9.36	9.27	9.16	9.02	8.85
6.40	9.00	8.98	8.92	8.84	8.74	8.61	8.46
6.60	8.59	8.56	8.51	8.44	8.34	8.23	8.09

H_{eq} (m)	$p_{traffic;max}$ (kPa)	$p_{traffic}$ (kPa) for pile arrangement (m^2)					
		0.5 x 0.5	1.0 x 1.0	1.5 x 1.5	2.0 x 2.0	2.5 x 2.5	3.0 x 3.0
6.80	8.20	8.17	8.13	8.06	7.97	7.87	7.74
7.00	7.83	7.81	7.77	7.71	7.63	7.53	7.42
7.20	7.49	7.47	7.43	7.38	7.31	7.22	7.12
7.40	7.17	7.16	7.12	7.07	7.00	6.92	6.83
7.60	6.88	6.86	6.83	6.78	6.72	6.65	6.56
7.80	6.60	6.58	6.55	6.51	6.45	6.39	6.31
8.00	6.33	6.32	6.29	6.25	6.20	6.14	6.07

$p_{traffic,max}$ is the maximum vertical load on the GR;

$p_{traffic}$ is the maximum average load on the GR. This value is used as the characteristic value of the traffic load p and to determine the value of κ for the reduction of the arching. Possibly, a permanent load may be added to this, so that the design load p becomes: $p = p_{traffic} + p_{permanent}$.

References

Standards / guidelines / instructions that are referenced to

[1] CUR 226, 2010. *Ontwerprichtlijn paalmatrassystemen (Design Guideline Piled Embankments)*, ISBN 978-90-376-0518-1 (in Dutch).

[2] CUR 226, 2016. *Ontwerprichtlijn paalmatrassystemen (Design Guideline Piled Embankments), updated version (in Dutch).*

[3] NEN-EN 1990+A1+A1/C2:2011 Eurocode 0: Basis of Structural Design, including the National Annex 2011.

[4] NEN-EN 1991-2+C1:2011 Eurocode 1: Actions on structures - Part 2: Traffic loads on bridges, including the National Annex 2011.

[5] NEN-EN 1992-1-1+C2:2011 Eurocode 2: Design of concrete structures - Part 1-1: General rules and rules for buildings, including the National Annex 2011.

[6] NEN-EN 1992-2+C1:2011 Eurocode 2: Eurocode 2: Design of concrete structures - Part 2: Concrete bridges - Design and detailing rules, including the National Annex 2011.

[7] NEN-EN 1997-1+C2:2012 Eurocode 7: Geotechnical Design– Part 1: General rules, including the National Annex 2012.

[8] NEN 9997-1+C1:2012 Eurocode 7: Geotechnical design of structures – Part 1: General rules – Composition of NEN-EN 1997-1:2005, NEN-EN 1997-1:2005/C1:2009 and the National Annex, complementary standard to NEN-EN 1997-1

[9] NEN-EN-ISO 22477-1, Geotechnisch onderzoek en beproeving – Beproeving van geotechnische constructies – Deel 1: Proefbelasting van palen door statische axiale belasting op druk.

[10] British Standard, BS 8006-1:2010 *'Code of practice for strengthened/reinforced soils and other fills'*, British Standards Institution, London.

[11] EBGEO, 2010. *Empfehlungen für den Entwurf und die Berechnung von Erdkörpern mit Bewehrungen aus Geokunststoffen*, vol. 2. Deutsche Gesellschaft für Geotechnik e.V. (DGGT, German Geotechnical Society), Auflage, ISBN 978-3-433-02950-3. Also available in English: Recommendations for Design and Analysis of Earth Structures using Geosynthetic Reinforcements EBGEO, 2011. ISBN 978-3-433-02983-1 and digital in English ISBN 978-3-433-60093-1.

[12] NEN-EN 13242+A1 *Aggregates for unbound and hydraulically bound materials for use in civil engineering work and road construction.*

Publications that are referenced to

[13] Bolton, M.D., 1986. *The strength and dilatancy of sands*, Géotechnique 36, No. 1, pp 65-78.

[14] Braunstorfinger, M. 1971. *Einfluss von Verkehrslasten gemäss DIN 1072 auf eingeerdete Rohre mit geringer Scheitelüberdeckung*, Rohre – Rohreleiungbau, Heft 4, August, 1971.

[15] Den Boogert, T.J.M., 2011. Piled Embankments with Geosynthetic Reinforcement, Numerical Analysis of Scale Model Tests. Master of Science thesis, Delft University of Technology.

[16] Heitz, Claas, 2006. B *odengewölbe unter ruhender und nichtruhender Belastung bei Berücksichtigung von Bewehrungseinlagen aus Geogittern*, November 2006, Kassel University Press GmbH, ISBN-10: 3-89958-250-0 en ISBN-13: 978-3-89958-250-5, PhD thesis.

[17] Niekerk, A.A. van, Molenaar, A.A.A. and Houben, L.J.M., 2002. *Effect of Material quality and compaction on the mechanical behaviour of base course materials and pavement performance*, in A.G. Correia and F.E.F. Branco (Eds.), Proceedings of the 6th international conference on the bearing capacity of roads and airfields, Lisbon, Portugal, 24-26 June 2002 (pp 1071-1079).

[18] Van Duijnen, P.G., Schweckendiek, T., Calle, E.O.F., Van Eekelen, S.J.M., 2015. Calibration of partial factors for basal reinforced piled embankments, In: Proceedings of ISGSR 2015 Risks, Rotterdam.

[19] Van Eekelen, Suzanne, Hein Jansen, Piet van Duijnen, Martin de Kant, Jan van Dalen, Marijn Brugman, Almer van der Stoel, Marco Peters, 2010. *The Dutch Design Guideline for Piled Embankments*, 9th International Conference on Geosynthetics – 9ICG, Brazil.

[20] Van Eekelen, S.J.M., Bezuijen, A., Lodder, H.J., van Tol, A.F., 2012b. *Model experiments on piled embankments*. Part II. Geotextiles and Geomembranes 32: 82-94 including its corrigendum: Van Eekelen, S.J.M., Bezuijen, A., Lodder, H.J., van Tol, A.F., 2012b2. Corrigendum to 'Model experiments on piled embankments. Part II' [Geotextiles and Geomembranes volume 32 (2012) pp 82-94]. Geotextiles and Geomembranes 35: 119.

[21] Van Eekelen, S.J.M., Bezuijen, van Tol, A.F., 2013. *An analytical model for arching in piled embankments*. Geotextiles and Geomembranes 39: 78-102.

[22] Van Eekelen, S.J.M., Bezuijen, van Tol, A.F., 2015. *Validation of analytical models for the design of basal reinforced piled embankments*. Geotextiles and Geomembranes 43: 56-81.

[23] Van Eekelen, S.J.M., 2015. *Basal Reinforced Piled Embankments*. PhD thesis TU Delft, ISBN 978-94-6203-826-4 (electronic) of ISBN 978-94-6203-825-7 (paper).

Other publications

[24] Abusharar, S.W., Zeng, J.J., Chen, B.G., Yin, J.H., 2009. *A simplified method for analysis of a piled embankment reinforced with geosynthetics*. Geotextiles and Geomembranes 27, 39-52.

[25] Alexiew, D., Vogel W., 2001. *Railroads on piled embankments in Germany: Milestone projects. In: Landmarks in Earth Reinforcement*. Swets and Zeitlinger, 2001, S. 185-190.

[26] Alexiew, Dimiter, 2004. *Piled embankments for railroads: short overview of methods and significant case studies*. Proceedings of the International Seminar on Geotechnics in Pavement and Railway Design and Construction, Gomes Correia and Loizos (eds), 2004 Millpress, Rotterdam, ISBN 90 5966 038 2, pp 181-192.

[27] Almeida, M.S.S., Ehrlich, M., Spotti, A.P., Marques, M.E.S., 2007. *Embankment supported on piles with biaxial geogrids*. Geotech. Eng., 160 (4), 185-192.

[28] Almeida, M.S.S., Marques, M.E.S., Almeida, M.C.F., Mendonca, M.B., 2008. *Performance on two low piled embankments with geogrids at Rio de Janeiro*. In: Proceedings of the First Pan American Geosynthetics Conference and Exhibition, Cancun, Mexico, 1285-1295.

[29] ASIRI, 2012. *Recommandations pour la conception, le dimensionnement, l'exécution et le contrôle de l'amélioration des sols de fondation par inclusions rigides*. ISBN: 978-2-85978-462-1.

[30] Bathurst R.J., Cai Z., 1994. *In-isolation Cyclic Load-Extension Behavior of Two Geogrids*. Geosynthetics International, Vol 1, No 1, pp 1-19.

[31] Bezuijen, A., van Eekelen, S.J.M., van Duijnen, P.G., 2010. *Piled embankments, influence of installation and type of loading*. In: Proceedings of the 9th International Conference on Geosynthetics, Brazil, 2010; 1921-1924.

[32] Blanc, M., Rault, G., Thorel, L., Almeida, M., 2013. *Centrifuge investigation of load transfer mechanisms in a granular mattress above a rigid inclusions network*. Geotextiles and Geomembranes 36: 92 - 105.

[33] Blanc, M., Thorel, L., Girout, R., Almeida, M., 2014. *Geosynthetic reinforcement of a granular load transfer platform above rigid inclusions; comparison between centrifuge testing and analytical modelling*. Geosynthetics International, Volume 21, Issue 1: 37-52.

[34] Blivet, J.C., Gourc, J.P., Giraud, H., 2002. *Design method for geosynthetics as reinforcement for embankment subjected to localized subsidence*. Proc. 7th International Conference on Geotextiles, pp 341-344.

[35] Briançon, L., Simon, B., 2012. *Performance of Pile-Supported Embankment over Soft Soil: Full-Scale Experiment*. J. Geotechn. Geoenviron. Eng. 2012.138:551-561.

[36] Britton, E.J., Naughton, P.J., 2008. *An experimental investigation of arching in piled embankments*. In: Proceedings of the 4th European Geosynthetics Congress EuroGeo 4. Edinburgh. Paper number 106.

[37] Britton, E.J., Naughton, P.J., 2010. *An experimental study to determine the location of the critical height in piled embankments.* In: Proceedings of the 9th Conference on Geosynthetics, Brazil, 1961-1964.

[38] Bussert, F., A.E.C. van der Stoel, Meyer, N. and A.P. de Lange, 2006. *Railway embankment on "high speed piles" – Design, installation and monitoring.* 8th International Conference on Geosynthetics (8ICG), Yokohama, Japan, September 2006.

[39] Cain, W., 1916. *Earth Pressure, Retaining Walls and Bins.* New York, John Wiley and Sons. Inc.

[40] Carlsson, B., 1987. *Reinforced soil, principles for calculation.* Terratema AB, Linköping (in Swedish).

[41] Casarin, C., 2011. Private communication. São Paulo, Brazil.

[42] CEN EN ISO 10319:2008 *Geosynthetics - Wide-width tensile test.* NEN, Delft, Netherlands.

[43] Chen, Y.M., Cao, W.P., Chen, R.P., 2008a. *An experimental investigation of soil arching within basal reinforced and unreinforced piled embankments.* Geotextiles and Geomembranes 26, 164-174.

[44] Chen, R.P., Chen, Y.M., Han, J., Xu, Z.Z., 2008b. *A theoretical solution for pile-supported embankments on soft soils under one-dimensional compression.* Can. Geotech. J. 45, 611-623.

[45] Collin, J.G., 2004. *Column supported embankment design considerations.* 52nd Annual Geotechnical Engineering Conference- University of Minnesota.

[46] Corbet, S.P., Horgan, G., 2010. *Introduction to international codes for reinforced soil design.* In: Proceedings of 9ICG, Brazil, pp. 225-231.

[47] Deb, K., 2010. *A mathematical model to study the soil arching effect in stone column-supported embankment resting on soft foundation soil.* Applied Mathematical Modelling 34 (2010), 3871-3883.

[48] Deb, K. Mohapatra, S.R., 2013. *Analysis of stone column-supported geosynthetic-reinforced embankments.* Applied Mathematical Modelling, Volume 37, Issue 5, 1 March 2013, 2943-2960.

[49] Demerdash, M.A., 1996. *An experimental study of piled embankments incorporating geosynthetic basal reinforcement.* PhD thesis, University of Newcastle upon Tyne, UK.

[50] Den Boogert, Th., van Duijnen, P.G., van Eekelen, S.J.M., 2012a. *Numerical analysis of geosynthetic reinforced piled embankment scale model tests.* Plaxis Bulletin 31, 12-17.

[51] Ellis, E., Aslam, R., 2009a. *Arching in piled embankments. Comparison of centrifuge tests and predictive methods, part 1 of 2.* Ground Engineering, June 2009, 34-38.

[52] Ellis, E., Aslam, R., 2009b. *Arching in piled embankments. Comparison of centrifuge tests and predictive methods, part 2 of 2*. Ground Engineering, July 2009, 28-31.

[53] Eskişar, T., Otani, J. and Hironaka, J., 2012. *Visualization of soil arching on reinforced embankment with rigid pile foundation using X-ray CT*. Geotextiles and Geomembranes 32: 44-54.

[54] Farag, G.S.F., 2008. *Lateral spreading in basal reinforced embankments supported by pile-like elements*. Schiftenreihe Getechnik, Heft 20, Universität Kassel, März 2008.

[55] Filz, G.M. and Smith, M.E., 2008. *Net Vertical Loads on Geosynthetic Reinforcement in Column-Supported Embankments*. Soil Improvement (GSP 172). Part of Geo-Denver 2007: New Peaks in Geotechnics. In: Proceedings of Sessions of Geo-Denver 2007.

[56] Filz, G., Sloan, J., McGuire, M., Collin, J., Smith, M., 2012. *Column-Supported Embankments: Settlement and Load Transfer*. Geotechnical Engineering State of the Art and Practice: 54-77. doi: 10.1061/9780784412138.0003.

[57] Filz, G., Sloan, J., 2013. *Load Distribution on Geosynthetic Reinforcement in Column-Supported Embankments*. In: Proceedings of Geo-Congress, California, 1829-1837.

[58] Fischer, L. 1960. *Theory and Practice of Shell Structures*. Berlin: Wilhelm Ernst and Sohn. 541 pages.

[59] Forsman, J., Honkala, A., Smura, M., 1999. *Hertsby case: a column stabilized and geotextile reinforced road embankment on soft subsoil. Dry mix method for deep soil stabilization*. H. Bredenberg, G. Holm, and B. B. Broms, eds., Balkema, Rotterdam, The Netherlands, 263–268.

[60] Forsman, J. 2001. *Geovahvistetutkimus, koerakenteiden loppuraportti 1996–2001 Georeinforcement-project, final report of test structures 1996–2001*. Tiehallinto, Helsinki, Finland (in Finnish).

[61] Gangakhedkar, Rutugandha, 2004. *Geosynthetic Reinforced Pile Supported Embankments*. Masters thesis of the University of Florida, University of Florida.

[62] Gavin, K.G.; O'Kelly, B.C., 2007. *Effect of friction fatigue on pile capacity in dense sand*. Journal of Geotechnical and Geoenvironmental engineering, January 2007.

[63] Giroud, J.P., Bonaparte, R, Beech, J.F., Gross, B.A., 1990. *Design of Soil Layer-Geosynthetics Systems Overlying Voids*. Geotextiles and Geomebranes, Elsvier, Vol. 9, pp 11-50.

[64] Guido, V.A., Kneuppel, Sweeney, M.A., 1987. *Plate loading test on geogrid reinforced earth slabs*. In: Proceedings Geosynthetics'87 Conference, New Orleans, 216–225.

[65] Habib, H.A.A., Brugman, M.H.A., Uijting, B.G.J., 2002. *Widening of Road N247 on a geogrid reinforced mattress on piles*. 7th International Conference on Geosynthetics, Nice.

[66] Halvordson, K.A., Plaut, R.H., Filz, G.M., 2010. *Analysis of geosynthetic reinforcement in pile-supported embankments*. Part II: 3D cable-net model. Geosynthetics International 17 (2), 68 - 76. ISSN: 1072-6349, E-ISSN: 1751-7613.

[67] Han, J., Gabr, M.A., 2002. *Numerical Analysis of Geosynthetic-Reinforced and Pile-Supported Earth Platforms over Soft Soil*. J. Geotechn. Geoenviron. Eng. January 2002; 44-53.

[68] Han, J., Bhandari, A., Wang, F., 2012. *DEM Analysis of Stresses and Deformations of Geogrid-Reinforced Embankments over Piles*. Int. J. Geomech. 2012.12:340-350.

[69] Handy, R. L., 1973. *The Igloo and the Natural Bridge as Ultimate Structures*. Arctic, Vol. 26, No. 4, 276-281.

[70] Handy, R.L., 1985, *The Arch in Soil Arching*. Journal of Geotechnical Engineering, Vol. Ill, No. 3, March, 1985. ©ASCE, ISSN 0733-9410/85/0003-0302/$01.00. Paper No. 19547.

[71] Haring, W., Profittlich, M., Hangen, H., 2008. *Reconstruction of the national road N210 Bergambacht to Krimpen a.d. IJssel, NL: design approach, construction experiences and measurement results*. In: Proceedings 4th European Geosynthetics Conference, September 2008, Edinburgh, UK.

[72] Hewlett, W.J., Randolph, M.F., 1988. *Analysis of piled embankments. Ground Engineering*, April 1988, Volume 22, Number 3, 12-18.

[73] Hong, W.P., Lee, J.H., Lee, K.W., 2007. *Load transfer by soil arching in pile-supported embankments*. Soils and Foundations Vol. 47, No. 5, 833-843.

[74] Hong, W.P., Lee, J., Hong, S., 2014. *Full-scale Tests on embankments founded on piled beams*. J. Geotech. Geoenviron. Eng. http://dx.doi.org/10.1061/(ASCE) GT.1943-5606.0001145.

[75] Horgan, Graham, 2006. *Code breaking (Revision of Section 8 of BS 8006 code of practice)*, Ground Engineering, October 2006.

[76] Horgan, G.J., and Sarsby, R.W., 2002. *The arching effect of soils over voids and piles incorporating geosynthetic reinforcement*. Geosynthetics, 7th ICG, Delmas, Gourc & Girard (eds) © Swets & Zeitlinger, Lisse ISBN 90 5809 523 1, 373-378.

[77] Huang, J., Han, j., Oztoprak, S., 2009. *Coupled Mechanical and Hydraulic Modeling of Geosynthetic-Reinforced Column-Supported Embankments*. J. Geotech. Geoenviron. Eng. 2009.135:1011-1021.

[78] Jenner, C.J., Austin, R.A., and Buckland, D., 1998. *Embankment support over piles using geogrids*. Proc. Sixth International Conf. Geosynthetics, 763-766.

[79] Jenck, O., Dias, D., Kastner, R., 2005, *Soft ground improvement by vertical rigid piles two-dimensional physical modelling and comparison with current design models*. Soils and Foundations, Vol. 45, No. 6, 15-30.

[80] Jenck, O., Dias, D., and Kastner, R., 2009. *Discrete element modelling of a granular platform supported by piles in soft soil – Validation on a small scale model test and comparison to a numerical analysis in a continuum*. Computers and Geotechnics 36: 917–927.

[81] Jones, B.M., Plaut, R.H., Filz, G.M., 2010. *Analysis of geosynthetic reinforcement in pile-supported embankments*. Part I: 3D plate model. Geosynthetics International 17 (2), 59e67. ISSN: 1072-6349, E-ISSN: 1751-7613.

[82] Jones, C.J.F.P., Lawson, C.R., Ayres, D.J., 1990. *Geotextile reinforced piled embankments*. In: Geotextiles, Geomembranes and Related Products, Den Hoedt (ed.), 1990, Balkema, Rotterdam, ISBN 90 6191 119 2, 155-160.

[83] Kempfert, H.G., Zaeske, D., Alexiew, D., 1999. *Interactions in reinforced bearing layers over partial supported underground*. In: Geotechnical Engineering for Transportation Infrastructure, Barends et al. (eds) © 1999 Balkema, Rotterdam, ISBN 90 5809 047 7.

[84] Kempfert, H.G., Göbel, C., Alexiew, D., Heitz, C., 2004. *German recommendations for reinforced embankments on pile-similar elements*. In: Proceedings of EuroGeo 3, Munich, 279-284.

[85] Kempton, G. Russell, D., Pierpoint, N.D. and Jones, C.J.F.P., 1998. *Two- and Three-Dimensional Numerical Analysis of the Performance of Piled Embankment*. Proceedings of the 6th International Conference on Geosynthetics, Atlanta.

[86] Lally, D. and Naughton, P.J., 2012. *An investigation of the arching mechanism in a geotechnical centrifuge*. In: Proceedings 5th European Geosynthetics Congress. Valencia. Vol 5. 363-367.

[87] Lawson, C.R., 1995. *Basal reinforced embankment practice in the United Kingdom, The practice of soil reinforcing in Europe*. Thomas Telford. London. 173-194.

[88] Le Hello, B., 2007. *Renforcement par geosynthetiques des remblais sur inclusions rigides, étude expérimentale en vraie grandeur et analyse numérique*. PhD thèses, l'université Grenoble I, (in French).

[89] Le Hello, B., Villard, P., 2009. *Embankments reinforced by piles and geosynthetics – Numerical and experimental studies with the transfer of load on the soil embankment*. Engineering Geology 106 (2009) 78–91.

[90] Liikennevirasto, 2012. *Geolujitetut maarakenteet, Tiegeotekniikan käsikirja, Liikenneviraston oppaita 2/2012*, ISBN 978-952-255-104-7. Finnish design guideline for geosynthetic reinforcement (in Finnish).

[91] Ling H.I., Mahri Y., Kawabaku T., 1998. *Tensile properties of geogrids under cyclic loading*. Journal of the Geotechnical en Geoenvironmental Engineering, Aug 1998, pp 782 – 787.

[92] Lodder, H.J., 2010. *Piled and reinforced embankments, Comparing scale model tests and theory*. Master of Science thesis, Delft University of Technology.

[93] Lodder, H.J., van Eekelen, S.J.M., Bezuijen, A., 2012. *The influence of subsoil reaction in a basal reinforced piled embankment*. In: Proceedings of Eurogeo5, Valencia. Volume 5.

[94] Love, J. and Milligan, G., 2003. *Design methods for basally reinforced pile-supported embankments over soft ground*. Ground Engineering. March 2003, 39-43.

[95] Low, B.K., Tang, S.K., and Chao, V., 1994. *Arching in piled embankments*. J. of Geo. Eng., ASCE, 120 (11), 1917-1938.

[96] Marston, A., Anderson, A.O., 1913. *The Theory of Loads on Pipes in Ditches and Tests of Cement and Clay Drain Tile and Sewer Pipe*. Bulletin No. 31, Engineering Experiment Station.

[97] McGuire, M.P., Filz, G.M., Almeida, M.S.S., 2009. *Load-Displacement Compatibility Analysis of a Low-Height Column-Supported Embankment*. In: Proceedings of IFCEE09, Florida, 225-232.

[98] McGuire, M., Sloan, J., Collin, J., Filz, G., 2012. *Critical Height of Column-Supported Embankments from Bench-Scale and Field-Scale Tests*. ISSMGE - TC 211 International Symposium on Ground Improvement IS-GI Brussels.

[99] McKelvey, J.A., 1994. *The Anatomy of Soil Arching*. Geotextiles and Geomembranes 13 (1994) 317-329.

[100] Nadukuru, S.S., Michalowski, R.L., 2012. *Arching in Distribution of active Load on Retaining Walls*. J. Geotechn. Geoenviron. Eng., May 2012. 575-584.

[101] Naughton, P., 2007. *The Significance of Critical Height in the Design of Piled Embankments*. Soil Improvement: 1-10. doi: 10.1061/40916(235)3.

[102] *Nordic guidelines for reinforced soils and fills (NGI), 2003*. Revision A was published in February 2004, and can be downloaded at www.sgf.net.

[103] Nunez, M.A., Briançon, L., Dias, D., 2013. *Analyses of a pile-supported embankment over soft clay: Full-scale experiment, analytical and numerical approaches*. Engineering Geology 153 (2013) 53-67.

[104] Oh, Y.I., Shin, E.C., 2007. *Reinforced and arching effect of geogrid-reinforced and pile-supported embankment on marine soft ground*. Marine Georesources and Geotechnology, 25, 97-118.

[105] Plaut, R.H., Filz, G.M., 2010. *Analysis of geosynthetic reinforcement in pilesupported embankments*. Part III: axisymmetric model. Geosynthetics International 17 (2), 77-85. ISSN: 1072-6349, E-ISSN: 1751-7613.

[106] Public Work Research Center, 2000. *Manual on Design and Execution of Reinforced Soil Method with Use of Geotextiles*. Second ed. Public Work Research Center, 248-256 (in Japanese).

[107] Rogbeck, Y., Gustavsson, S., Södergren, I. Lindquist, D., 1998. *Reinforced Piled Embankments in Sweden - Design Aspects*. In: Proceedings of the Sixth International Conference on Geosynthetics, 755-762.

[108] Russell, D., Pierpoint, N., 1997. *An assessment of design methods for piled embankments*. Ground Engineering, November 1997, 39-44.

[109] SINTEF, 2002. *A computer program for designing reinforced embankments*. In: Proceedings of the 7 ICG, Nice 2002, France, vol. 1, 201–204.

[110] Slaats, H. and van der Stoel, A.E.C., 2009. *Validation of numerical model components of LTP by means of experimental data*. 17th International Conference on Soil Mechanics and Geotechnical Engineering, Alexandria, Egypt, 5-9 October 2009.

[111] Sloan, J.A., 2011. *Column-supported embankments: full-scale tests and design recommendations*. PhD thesis, Virginia Polytechnic Institute and State University, Blacksburg, VA.

[112] Spotti, A.P., 2006. *Monitoring results of a piled embankment with geogrids* (in Portuguese). ScD thesis. COPPE/UFRJ, Rio de Janeiro, Brazil.

[113] Stewart, M.E. and Filz, G., 2005. *Influence of Clay Compressibility on Geosynthetic Loads in Bridging Layers for Column-Supported Embankments*. In: Proceedings of Geo-Frontiers 2005, USA, GSP 131 Contemporary Issues in Foundation Engineering.

[114] Svanø, G., Ilstad, T., Eiksund, G., Want, A., 2000. *Alternative calculation principle for design of piled embankments with base reinforcement*. In: Proceedings of the 4th GIGS in Helsinki.

[115] Terzaghi, K., 1943. *Theoretical Soil Mechanics*. John Wiley and Sons, New York.

[116] Tonks, D., Hillier, R., 1998. *Assessmen revisited. Further discussion on "An assessment of design methods for piled embankments" by D. Russell and N.D. Pierpoint*. Ground Engineering November 1997. Ground Engineering, June 1998, pp 46-50.

[117] Van der Peet, T.C., 2014. *Arching in basal reinforced piled embankments, numerical validation of the Concentric Arches model*. MSc thesis, Delft University of Technology, Delft, the Netherlands.

[118] Van der Peet, T.C., van Eekelen, S.J.M., 2014. *3D numerical analysis of basal reinforced piled embankments*. In: Proceedings of IGS10, September 2014, Berlin, Germany. Paper no. 112.

[119] Van der Stoel, A.E.C., J.M. Klaver, A.T. Balder and A.P. de Lange, 2006. *Numerical design, installation and monitoring of a load transfer platform (LTP) for a railway embankment near Rotterdam*. Sixth European Conference on Numerical Methods in Geotechnical Engineering, Graz University Of Technology, September 2006.

[120] Van der Stoel, A.E.C., D. Vink, R.W. Ravensbergen and M. de Hertog, 2007. *Design and execution of an integrated LTP and gabions system*. In: proceedings of XIV European Conference on Soil Mechanics and Geotechnical Engineering. Madrid September 2007.

[121] Van der Stoel, A.E.C., Brok, C., De Lange, A.P., van Duijnen, P.G., 2010. *Construction of the first railroad widening in the Netherlands on a Load Transfer Platform (LTP)*. In: Proceedings of 9 ICG, Brazil, 1969-1972.

[122] Van Duijnen, P.G., van Eekelen, S.J.M., van der Stoel, A.E.C., 2010. *Monitoring of a Railway Piled Embankment*. In: Proceedings of 9 ICG, Brazil, 1461-1464.

[123] Van Eekelen, S.J.M., Bezuijen, A. Oung, O., 2003. *Arching in piled embankments; experiments and design calculations*. In: Proceedings of Foundations: Innovations, observations, design and practice, 885-894.

[124] Van Eekelen, S.J.M., Van, M.A., Bezuijen, A., 2007. *The Kyotoroad, a full-scale test. Measurements and calculations*. In: Proceedings of ECSMGE 2007, Madrid, Spain, 1533-1538.

[125] Van Eekelen, S.J.M. en Bezuijen, A., 2008. *Considering the basic starting points of the design of piled embankments in the British Standard BS8006*. In: Proceedings of EuroGeo4, paper number 315, September 2008, Edinburgh, Scotland.

[126] Van Eekelen, S.J.M., A. Bezuijen, P. van Duijnen and H.L. Jansen, 2009. *Piled embankments using geosynthetic reinforcement in the Netherlands: design, monitoring and evaluation*. Proceedings of the 17th International Conference on Soil Mechanics and Geotechnical Engineering, M. Hamza et al. (Eds.), 2009 IOS Press., pp 1690-1693.

[127] Van Eekelen, S.J.M., Bezuijen, A., Alexiew, D., 2010a. *The Kyoto Road Piled Embankment: 31/2 Years of Measurements*. In: Proceedings of 9 ICG, Brazil, 1941-1944.

[128] Van Eekelen, S.J.M., Bezuijen, A., van Tol, A.F. (2011a). *Analysis and modification of the British Standard BS8006 for the design of piled embankments. Geotextiles and geomembranes 29: 345 - 359*

[129] Van Eekelen, S.J.M., Bezuijen, A., Lodder, H.J., van Tol, A.F. (2012a). *Model experiments on piled embankments. Part 1.Geotextiles and geomembranes 32: 69 - 81.*

[130] Van Eekelen, S.J.M., Lodder, H.J., Bezuijen, A., 2011b. *Load distribution on the geosynthetic reinforcement within a piled embankment*. In: Proceedings of ICSMGE 2011, Athens, 1137-1142.

[131] Van Eekelen, S.J.M., Bezuijen, A., van Duijnen, P.G., 2012c. *Does a piled embankment 'feel' the passage of a heavy truck? High frequency field measurements*. In: Proceedings of the 5th European Geosynthetics Congress EuroGeo 5. Valencia. Digital version volume 5: 162-166.

[132] Van Eekelen, S.J.M., Bezuijen, A., 2012a. *Inversed triangular load distribution in a piled embankment, 3D model experiments, field tests, numerical analysis and consequences*. EuroGeo5, Valencia, Spain.

[133] Van Eekelen, S.J.M., Bezuijen, A. 2012b. *Model experiments on geosynthetic reinforced piled embankments, 3D test series*. In: Proceedings of Eurofuge, Delft.

[134] Van Eekelen, S.J.M., Bezuijen, A. 2012c. *Basal reinforced piled embankments in the Netherlands, Field studies and laboratory tests*. ISSMGE - TC 211 International Symposium on Ground Improvement IS-GI Brussels.

[135] Van Eekelen, S.J.M. and Bezuijen, A. 2013a. *Equilibrium models for arching in basal reinforced piled embankments*. In: Proceedings of 18th ICSMGE, Paris. 1267 – 1270.

[136] Van Eekelen, S.J.M., Bezuijen, A., 2013b, *Dutch research on piled embankments*. In: Proceedings of GeoCongress, California, March 2013, 1838-1847.

[137] Van Eekelen, S.J.M., Almeida, M.S.S., Bezuijen, A., 2014. *European analytical calculations compared with a full-scale Brazilian piled embankment*. In: Proceedings of 10ICG, Berlin, Germany. Paper no. 127.

[138] Van Eekelen, S.J.M., Bezuijen, A., 2014. *Is 1 + 1 = 2? Results of 3D model experiments on piled embankments*. In: Proceedings of 10ICG, Berlin, Germany. Paper no. 128.

[139] Van Eekelen, S.J.M., Bezuijen, A., van Tol, A.F., 2015b. *Axial pile forces in piled embankments, field measurements*. In: Proceedings of XVI ECSMGE, Edinburgh.

[140] Van Niekerk, A.A., Molenaar, A.A.A., Houben, L.J.M., 2002. *Effect of Material Quality and Compaction on the Mechanical Behaviour of Base Course Materials and Pavement Performance. In: Proceedings of the 6th international conference on the bearing capacity of roads and airfields*. Lisbon, Portugal, June 2002: 1071-179. Swets and Zeitlinger BV.

[141] Vermeer, P.A., Punlor, A., Ruse, N., 2001. *Arching effects behind a soldier pile wall*. Computers and Geotechnics 28 (2001) 379–396.

[142] Vollmert, L., Kahl, M., Giegerich, G., Meyer, N., 2007. *In-situ verification of an extended calculation method for geogrid reinforced load distribution platforms on pile foundations*. In: Proceedings of ECSGE 2007, Madrid, Volume 3, 1573 - 1578.

[143] Weihrauch, S., Oehrlein, S., Vollmert, L., 2010. *Baugrundverbesserungsmassnahmen in der HafenCity Hamburg am Beispiel des Stellvertreterprojektes Hongkongstrasse*. Bautechnik. Volume 87, issue 10: 655-659 (in German).

[144] Weihrauch, S., Oehrlein, S., Vollmert, L., 2013. *Subgrade improvement measures fort the main rescue roads in the urban redevelopment area HafenCity in Hamburg*. In: Proceedings of 18th ICSMGE, Paris.

[145] Wood, H., Horgan, G., Pedley, M., 2003. A63 *Selby bypass – design and construction for a 1.6 km geosynthetic reinforced piled embankment.* Proc. EuroGeo 3, pp 299-304.

[146] Zaeske, D., 2001. *Zur Wirkungsweise von unbewehrten und bewehrten mineralischen Tragschichten über pfahlartigen Gründungselementen.* Schriftenreihe Geotechnik, Uni Kassel, Heft 10, February 2001 (in German).

[147] Zaeske, D. and Kempfert, H.-G., 2002. *Berechnung und Wirkungsweise von unbewehrten und bewehrten mineralischen Tragschichten auf punkt- und linienförmigen Traggliedern.* Bauingenieur Band 77, Februar 2002.

[148] Zhang, L., Zhao, M., Hu, Y., Zhao, H., Chen, B., 2012. *Semi-analytical solutions for geosynthetic-reinforced and pile-supported embankment.* Computers and Geotechnics, Volume 44, June 2012, 167-175, ISSN 0266-352X, 10.1016/j. compgeo.2012.04.001.

[149] Zhuang, Y., Wang, K.Y., Liu, H.L., 2014. *A simplified model to analyze the reinforced piled embankments.* Geotextiles and Geomembranes, Volume 42, Issue 2, April 2014, 154-165.